D1306131

the visionary package

Also by the same authors:

Branding @ the digital age (eds), Palgrave Macmillan

Also by Herbert Meyers:

The Marketer's Guide to Successful Package Design
  (co-author Murray J. Lubliner, McGraw-Hill)

the
visionary
package

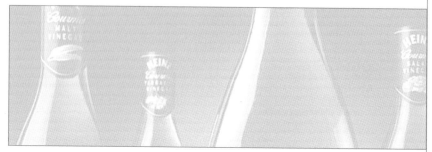

using packaging to
build effective brands

Herbert Meyers and **Richard Gerstman**

palgrave
macmillan

First published 2005 by
PALGRAVE MACMILLAN
Houndmills, Basingstoke, Hampshire RG21 6XS and
175 Fifth Avenue, New York, N.Y. 10010
Companies and representatives throughout the world

PALGRAVE MACMILLAN is the global academic imprint of the Palgrave Macmillan division of St. Martin's Press, LLC and of Palgrave Macmillan Ltd. Macmillan® is a registered trademark in the United States, United Kingdom and other countries. Palgrave is a registered trademark in the European Union and other countries.

ISBN 1–4039–0677–7

This book is printed on paper suitable for recycling and made from fully managed and sustained forest sources.

A catalogue record for this book is available from the British Library.

A catalog record for this book is available from the Library of Congress.

10 9 8 7 6 5 4 3 2 1
14 13 12 11 10 09 08 07 06 05

Printed in China

Contents

contents

About the authors

Herbert Meyers and Richard Gerstman are the retired founding partners of Gerstman+Meyers (now Interbrand), a global design consultancy.

Herb Meyers was born in Germany and came to the U.S. in 1939. After serving as an interpreter in the United States Army Air Corps during World War II, he studied design at the Pratt Institute from where he holds a Bachelor of Fine Arts degree.

Richard Gerstman is the chairman of Interbrand U.S. He graduated as an industrial designer from the University of Cincinnati, then worked in Norway and Sweden, countries that were in the forefront of design at the time.

Both Herb Meyers and Richard Gerstman realized early in their careers that the package could become an influential item in branding and promoting goods for the marketplace, and were interested in the appeal and potential of well-designed packaging.

In 1970, Herb and Richard met and founded Gerstman+Meyers, Starting as a small group, they personally did everything from designing to supervising, client contact, new business development, and writing proposals. As their business grew in reputation and number of clients, they pioneered the recommendation of designs based on market conditions and aesthetics. Together with their staff, they provided services in brand identity, structural and graphic package design, corporate identity and environmental interiors for major corporations in the U.S., Canada, South America, Europe and the Far East.

While passionately creating branding and package design for single products and extensive brand lines, Herb and Richard never relaxed their focus of building a reputation of excellence and reliability. Their clients included such companies as Johnson & Johnson, Procter & Gamble, AT&T, Heinz, Kellogg's, Ralston-Purina, Black & Decker, BASF, Bayer, General Motors, Pepsi-Cola, Omni Hotels and Maxell.

The company was honored with over 300 design awards. Richard Gerstman himself owns several design and utility patents. As past president of the Package Design Council International (PDC), Herb

Meyers was the first person ever to be awarded the organization's PDC Award for Lifetime Packaging Excellence and Leadership.

Both authors have been frequent writers and lecturers on the subject of branding and packaging and recently co-authored the book *Branding @ the digital age* (Palgrave Macmillan). Herb Meyers also co-authored *The Marketer's Guide to Successful Package Design* (McGraw-Hill) with Murray J. Lubliner.

In 1996, Gerstman+Meyers became part of the Interbrand Group. Interbrand, one of the world's leading brand consultancies, specializes in brand strategy and consumer research, as well as packaging design, naming and brand valuation for clients throughout the world.

Acknowledgements

We are grateful to the many individuals who contributed their time and attention in helping to provide us with information for this book.

We would particularly like to thank Chuck Brymer, group CEO of Interbrand, who encouraged and supported the writing of this book, and Jeff Swystun, global director of knowledge & innovation at Interbrand, who reviewed many parts of the manuscript and made helpful suggestions.

This book required a substantial amount of research and interviews, and many of the individuals whom we interviewed put aside a large chunk of their busy schedules for us. We are especially thankful to a long list of contributors that includes:

Bob Anderson, vice-president private label products, Wal-Mart

Rowland Archer, chief technology officer & co-founder, HAHT Commerce

Kevin Ashton, executive director, Auto-ID Center, MIT Media Laboratories

Richard Cantwell, vice-president Auto-ID, Gillette Corporation

Lee Carpenter, CEO, Design Forum

Gayle Christensen, director of marketing & global brand management, FedEx

Howard Clabo, senior communication specialist, FedEx

John Clarke, director of group technology, Tesco

Pam DeCesare, director, packaging & brand design, Kraft Foods

Mona Doyle, president, Consumer Research Network

Bob Gallagher, public relations manager, HAHT Commerce

Eric F. Greenberg, Principal Attorney, Eric F. Greenberg, P.C.

Art Herstol, associate design director, beauty care, Procter & Gamble

Maggie Jackson, columnist and writer

Richard Lewis, vice-president account services, TBWA Worldwide

John Nottingham, co-president, Nottingham-Spirk Design Associates

Brian Perkins, worldwide chairman, Consumer Pharmaceuticals & Nutritional Group, Johnson & Johnson

Marc Rosen, president, Marc A. Rosen Associates

acknowledgements

Sanjay Sarma, associate professor & research director, Auto-ID Center, MIT Media Laboratories

Perry Seelert, vice-president product strategy, Damon Worldwide

Brian Sharoff, president, Private Label Manufacturers Association

Bob Swientek, editor-in-chief, *BrandPackaging* Magazine

J.P. Terry, CEO, BrandWizard Technologies

Dane Twining, director of public relations, Private Label Manufacturers Association

Lars Wallentin, asssistant vice-president, marketing communications & strategic design, Nestlé

Elliott Young, chairman, Perception Research Services International

Stan Zelesnik, director of education, Institute of Packaging Professionals

Sharon Gordon, director of network resources at Omnicom Group was very helpful in providing us with names of some of the people to contact for interviews, as was Phyllis Kim, vice-president, Fleishman-Hillard in St. Louis.

We are also grateful to Interbrand offices around the world for contributing information on packaging case histories in their regions. Special thanks to Terry Oliver, president and CEO of the Asia Pacific offices, Jeremy Scholfield, director/brand identity at the London office, Patrick Ashe, associate director/graphics at the Singapore office, Facundo Bertranou, creative manager at the Buenos Aires office, and Kathy Hoopes, director private brands at the Cincinnati office.

Special commendations go to Chris Campbell, creative director and Chris Armstrong, senior designer at Interbrand's Toronto office for art directing the cover graphics. Heather Botjer took many of the fine photographs. Two long-time colleagues at Interbrand's New York office, Sal D'Orio, director of production services, and Miguel Rivera, production manager, contributed the technical expertise for digitizing and coordinating the photographs that have been gathered for the book. Lisa Marsala, director, marketing/PR at the New York office helped us immensely with publicity information.

No efforts in connection with this book should ignore the patience of our beloved spouses for sharing our time and attention with the book's creation. We believe that they will ultimately participate in our pride in presenting *The Visionary Package* to our readers.

Introduction
What is visionary packaging?

"The wave of the future is coming and there is no fighting it", wrote Anne Morrow Lindbergh in 1940. No better description could apply to our lives today.

What will the future bring? None of us really knows for sure. But if you could predict future lifestyles, shopping patterns, stores, technology and how your competition will think in the future, you would certainly breathe more easily and market your products with less worry.

Since none of us are mind-readers, we need to look for other ways of enduring the test of time with a vision for the future. You can define vision as having the ability to anticipate. Making provision for future events. Having foresight.

Vision implies a promise. A promise to achieve things that no one else has done or thought of before. A promise to go beyond the ordinary. A promise to be a leader.

But is there such a thing as visionary packaging? With so many packages around, what defines a package as being visionary? What makes one package stand out as being visionary while others are not?

Do some packages really anticipate future trends, or lifestyles, or shopping patterns, or technology? How can you foresee what lies ahead?

You need the vision

There's an expression: "You can observe a lot by just watching." Making a package visionary is not that simple. Vision is not an element to be included in your package design criteria, and you certainly cannot ask your designer to "design a visionary package".

But don't be discouraged. All is not lost. What you can do is have the information and foresight to develop criteria that will make your package development so well targeted that your package may join the exclusive club of visionary packages.

There have been a few such packages in the market that we can identify as being visionary. Let's single out some of them: Heinz Ketchup, Breyers Ice Cream, Campbell's Soup, L'eggs Pantyhose, Dunkin Donuts, Coca-Cola, Tanqueray Gin, Absolut Vodka, Bayer Aspirin, Marlboro Cigarettes. There are more, but let's do with these few.

Why do we feel that these packages have the distinction of being visionary packages? What makes them different from "ordinary" packages?

Take a close look. Most of them are not among the most wildly creative designs. They are simple, "straightforward" designs. But each has an element that you can identify as being memorable and owning a special personality.

Did you notice the long neck of Heinz Ketchup bottles, the black background of Breyers Ice Cream, the red and white colors of Campbell's Soup labels, the apothecary-like shape of Absolut Vodka?

What is so special about these features? What is the common thread that makes them visionary?

The common thread is that their marketers correctly anticipated the future based on specific, clearly defined criteria, a solid understanding of consumer needs and the ability to translate these into visual comprehensible units. Certainly, the original designers of these packages did not approach their task with the conscious intention of being "visionary". What makes these packages visionary is that, being based on solid marketing criteria and precisely targeted positioning instead of on marketers' or designers' whims or current popular trends, they have withstood the flow of time, even when modified repeatedly, to adhere to the changes in market conditions.

Thus, all succeeded in surviving fierce competitive, environmental, technological and lifestyle changes, some for more than half a century.

This book, *The Visionary Package*, focuses on the consumer and what your market is now, what your market could be, and how the market will change. You need this vision of what your market could be before the others beat you to it. You need a vision of how your product's packaging can break through the clutter now and remain there in the future. Because against your competition in the supermarket, the home center, the retail chain, or even the Internet, the packages with vision will be the winners.

But, to accomplish this in the future, one critical attitude that is often overlooked must make amends. It goes back all the way to your college marketing classes, when your professor handed you the sacred pillars of marketing – the four "Ps"– Product, Price, Place, Promotion.

What's wrong with the four "Ps", you may wonder? Well, nothing is wrong with the four "Ps", but we ask: Should we limit the pillars of marketing to just four "Ps"? How many "Ps" are there really?

The dilemma of the "Ps"

Professor Phillip Kotler, the well-respected authority on marketing, acknowledges the legitimacy of the four "Ps", but, importantly, he adds another "P". The fifth "P", Kotler suggests, should be Positioning.

To make his point, Kotler, in one of his writings, refers to the positioning strategy for L'eggs, a pantyhose brand, developed by Hanes several years ago especially for supermarket distribution. The pantyhose was packaged in an ingeniously designed egg-shaped plastic container, mirroring the brand name, that we will discuss in greater detail later.

Although Kotler does acknowledge the package, he attributes the success of the brand entirely to its positioning. Coming from one of our most revered marketing gurus, this interpretation of positioning cannot be ignored, But we believe that, in doing so, Kotler minimizes another "P"– Packaging, without which the L'eggs brand success would have been unlikely.

There is no question that the positioning of a brand is certainly paramount and that it precedes everything that follows it. But, in retailing, the lack of appropriately designed and executed packaging makes the cleverest positioning strategy impotent.

In the store, it is the package that communicates the positioning of the brand. It is the package that translates the brand's position into a visual entity. It's the package that conveys the product's benefits to the consumer. It's the package that drives the final purchase decision at the point of sale.

The package is an indispensable means of communicating everything about the brand. Certainly, without great brand positioning there will not be great packaging. But conversely, without the follow-through of the package, even the most ingenious positioning concept will fizzle out. For that reason, we believe that, in retail marketing, a sixth "P" – Packaging – is as critical as all the other "Ps", if not more so.

So, welcome the sixth "P" in marketing: Packaging.

Exploring the package

We believe that packaging in the next decade will be especially important, because we anticipate that shopping will experience revolutionary changes resulting from limitless conceptual opportunities for making the shopping environment more exciting.

Our book will explore the role that, we believe, packaging will play in future developments of streamlining the shopping experience.

Referring to some visionary packages of the past and present, we will discuss information and guidelines for marketers, retailers, designers, producers, and anyone who is interested in how brands and packaging will be affected by retail developments and the Internet in the next five or ten years.

The emphasis of the book will revolve around opportunities for promoting brands and products through packaging and brand identity. The book will explore opportunities that packaging can offer to the world of online and offline commerce and how marketers can take advantage of opportunities that they may not have considered before.

Our discussions will include:

- how packaging grew from a primitive beginning to a primary marketing tool
- how packaging affects emotions
- how the continued influence of branding will impact on packaging
- how packaging is vital to creating brand loyalty
- how package shapes and graphics affect the shopping experience
- how management styles will influence future packaging
- how the rise of private labels will impact on the retail trade
- how changing lifestyles will affect package design
- how marketers view brand and package designers
- how the digital age will affect packaging
- how to exploit promotional opportunities through packaging
- how clients and advertising agencies must interface with design agencies in their combined effort to create packaging that will be visionary

By discussing, in some detail, the conditions that led to visionary packages, we hope to help marketers and designers to consider their own circumstances and develop packages that may become the visionary packages of tomorrow.

section one

packaging
from horsebacks to hypermarkets

Chapter 1
The ubiquitous package

If someone asked you to think of an object that most often impacts on your daily tasks, what would it be? Would it be packaging?

Not very likely.

But think about it. Can you imagine going through even a single day without interacting with numerous objects that are packaged? Unlikely! In today's environment virtually everything is packaged. In many cases, the primary package, such as a glass bottle or a sensitive instrument, may even be double or triple packaged.

Why don't we start with your morning routine. You get up to prepare for the day. You start with pulling some dental floss out of a snap-open container, wash and style your hair, apply deodorant with your roll-on dispenser, spray on some lotion from an aerosol container or splash on a few drops of perfume from your favorite bottle.

And don't forget the tube of toothpaste and the bottle of mouthwash to freshen your breath.

Now, you walk into the kitchen, open the carton of orange juice for breakfast, shake a vitamin pill from its bottle, remove the paper wrapping from a loaf of bread or a package of muffins, scoop butter from its paper tub and some grape jelly from a glass jar, take the coffee filter paper from its carton, scoop coffee from its can, tear open a packet of sugar substitute… and so it goes on and on.

And you haven't even left home yet.

A starring role

You can't escape the ubiquitous package – it is everywhere. We may not pay much attention to it, and we certainly do not give packages much intellectual thought, but they are there at every step of our lives, morning, noon and night. Life without packaging is unthinkable these days.

How many things can you think of that play such a starring role in your daily life?

If packaging makes such a conspicuous contribution to our daily activities, what makes it so? When did it gain such vital importance in our private and business lives? What benefits does it promise for being such an intimate and central component of our lives? In what way does packaging contribute to our existence? Our lifestyles?

It deserves examination. So, let's explore it.

What's in a name?

Searching to disseminate the meaning of the word "packaging" leads us first to our always informative *Webster's New World Dictionary*. There you will find packaging described as "a wrapped or boxed thing or group of things".

If that sounds rather mundane, let's look further. *Webster's Thesaurus* does a more thorough job of expanding on the word "package" by rattling off a mind-boggling list of identities that include packet, parcel, bundle, kit, pack, batch, grip, bag, box, carton, crate, tin, sack, bottle… and the list goes on.

But language differences often lead to different conclusions. What is called a "package" in the US is called a "pack" in the UK. If we turn again to *Webster's Dictionary*, we find that "pack" is described in a lengthy list of interpretations, among them this rather intriguing one:

> a large bundle of things wrapped or tied for carrying on the back of an animal.

Now, if this may sound a bit exotic to you, it was, in fact, the beginning of packaging.

In today's commercial environment, we are used to thinking of packaging as a means of marketing products that are enrobed in paper, metal, glass or plastic to protect the contents from spoilage, contamination or damage, especially when hauled over long distances by plane, ship, rail or truck.

But that's today's living environment. It is important for us to remember that packaging, like most segments of our contemporary life, had its origin a long time ago. In fact, packaging food and a variety of objects has a history dating back thousands of years.

So come along for a short history of packaging.

Chapter 2
The evolution of packaging

Once upon a time...

There was a time when pack animals were a common means of transportation. A few thousand years ago, packages were primitive bundles, baskets or earthen containers that were created to hold and transport food, beverages or objects valuable to the members of ancient communities. Numerous antiquarian discoveries of such containers have been made all over the world, especially along the ancient silk roads that led from Asia through Persia, Phoenicia and Mesopotamia (now roughly the areas covered by Iran, Iraq and Syria) to Turkey, the Middle East, Africa and, eventually, to Southern and Central Europe.

The fact is that containers, more than any other evidence of human behavior, have been both witnesses and influencers of the evolution of lifestyles of the human race throughout the ages, from prehistoric times until today.

Ancient trailblazers

Initially, the ancient containers, ranging from simple woven baskets to elaborately structured and decorated bowls, jars, bottles and carafes, were created simply for the utilitarian purpose of holding and transporting food, beverages and condiments. Later, containers were created to store wine, jewelry, perfume and a wide variety of personal possessions. In time, many were decorated elaborately by their owners or artisans to please the eye. They can be admired in museums throughout the world where they are exhibited as treasured pieces of sculptural and graphic beauty or to reflect their genesis and purpose in terms of historical or anthropological significance.

Examples include the beautifully painted vases from the fourth mil-

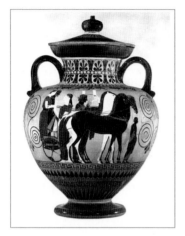

lennium BC at the Egyptian Museum in Cairo and the superbly decorated terracotta amphorae of the Greek period at the National Archeological Museum in Athens. We marvel at the artistic capabilities of ancient artisans resulting in the magnificent bronze wine vessels of ancient China and, at the National Museum of Anthropology and Archeology in Lima, Peru, the simplicity of primeval vessels, sculptured by people with no artistic training but an instinctive sense of design.

The Greeks stored food and beverages in elaborately decorated containers, such as this amphora with cover, c.540 B.C.

The Metropolitan Museum of Art, Rogers Fund. 1917

While many of these containers were used as household utilities, still others were utilized to store items of religious and ritual significance. Some took on special connotations among ancient communities who believed that certain containers held magic powers.

Although primarily functional and not as yet visionary, it's worth remembering that these early containers, whether primitive or artistically magnificent, were truly the precursors of our modern use of packaging. In fact, they initiated an evolutionary trend, suggesting that the container's importance rivaled its contents.

The changing role of packaging

While the archeological, historical and visual significance of these early containers is critical for understanding our history and lifestyles through the ages, containers played quite a different role in ancient times than that which we associate the purpose of containers with today.

For the earliest inhabitants on earth, whose subsistence relied on hunting and fishing to feed themselves and their families, nature provided shells and animal organs to preserve what they could not consume immediately. Gradually, generation by generation, our ancestors developed skills enabling them to store food in hollowed-out logs, fashion animal furs and skins into bags for food preservation and weave grasses and reeds into baskets to hold various objects.

The ancient skills of combining storage and art is evident in this Assyrian storage jar, from Zawiyeh, Azerbaijan, 7th century B.C.

By ancient standards, all were, in effect, packages.

As innovations accelerated and the skill of creating things improved, people discovered that pottery made of clay was better able to provide repositories for preserving food and drinks and which were more practical for holding personal possessions. Judging by the numerous examples unearthed, clay containers were used abundantly for those functions all over the world.

Thanks to the ancient potters, the fascination with pottery has remained alive to our present time and is still practiced as enthusiastically as ever by amateurs and skilled craftsmen alike. The town of Hagi on the island of Honshu, Japan, exemplifies a distinguished location for hosting a busy contingent of potters whose creative output of clay pottery is distributed throughout the world. In the South of France, the town of Valauris, renowned for its pottery industry, gained fame when Pablo Picasso's creative genius produced over 3000 pieces of magnificent pottery there.

Although some of the early clay containers had covers, they were more often open jars and bowls, capable of protecting their contents

Picasso created magnificently decorated ceramics that mirrored ancient pottery, such as this limited edition Picasso authentic replica by Galerie Madoura

temporarily, but not for any length of time, depending on the type of contents and nature's elements in the region in which they were used.

Today, of course, the primary purpose of containers is to provide long-lasting protective benefits. Thus, the characteristics of contemporary containers are that most of them fully enclose their contents, offering protection for days, weeks, or even years, depending on the type of product and product use. Many of them, having initially been conceived with jealously guarded trade names, have come to be household generics for identifying certain packaging forms, such as folding carton, blister pack, shrink-wrap and cluster pack.

Yet, with all this emphasis on packaging these days, packages that are truly visionary, that is, those that were the first to anticipate some kind of ergonomic or visual human need, are few and far between.

The first visionary package

The honor of being one of the first truly "visionary packages" belongs to containers created over 2000 years ago and discovered at an unlikely venue.

In 1947, a group of Bedouins in the Judean desert accidentally came across a number of long-abandoned caves where they found a large number of earthen jars. On removing the tops of the jars, they discovered that they contained fragile scrolls of parchment or leather, with mysterious scripts on them. As scientists excitedly descended upon the unusual discovery location, the jars and their contents became one of most important historic discoveries ever. What these Bedouins had hit upon was the location of the ancient community of Qumran, dating from the third century B.C.E., the time of the Maccabees.

The contents of these "packages", the ancient scrolls, known today as the "Dead Sea Scrolls" because of the location of the caves near the Dead Sea, contained detailed information about the life and the people who lived in the area at

A primitive Roman period tribe in the Judean desert created containers that kept fragile scrolls of parchment and leather from deteriorating for 2000 years

Photo © The Israel Museum, Jerusalem

that time. The scrolls, written in Aramaic, are one of the oldest cohesive religious and historical documents ever discovered.

The scrolls have been subjects of intense studies, an activity that would not have been possible except for the existence of containers that were able to protect them against atmospheric decay for over 2000 years. By creating jars made of earthen material, with tightly fitting covers that were able to resist the scourge of time and weather, for the sole purpose of protecting their contents from deterioration, a reclusive desert tribe created the first truly visionary package.

Exploiting the vision

While generations of early tribes produced urns, bowls and pottery with various degrees of sophistication, other communities developed more sophisticated means of producing containers. Most prominent is the early development and evolution of glass.

Attempting to be a replacement of earthen pottery, a combination of limestone, sand, soda and silica was melted together and molded into pottery-like, semi-transparent glass containers and gradually into smaller objects, such as cups and bowls. There is evidence of heavy demand among the Egyptian royalty, more than 3000 years ago, for glass bowls and bottles in a variety of translucent colors and distinctive shapes. These Egyptian luminaries loved the good life and fancied bottles for perfume that, it is reported, they applied liberally on themselves.

From the primitive methods of early glass production, glassmaking technology gradually grew into sophisticated glassmaking skills. When the Phoenicians invented glassblowing, the ability to create a wide variety of glass shapes led to a thriving glass industry that gradually migrated from the Middle East to Europe. Romans, in particular, treasured beautiful, colorful glass objects. Produced by artisans, their glassmaking techniques were shrouded in secrecy and were actually lost for a period of time to succeeding generations.

When the growing appetite on the part of European royalty, especially in Italy and France, led to a revival of glassmaking, highly creative craftsmen, especially on the Venetian island of Murano, developed glassblowing expertise that was considered to be so precious at that time that any glassmaker who wanted to leave the island risked the death penalty.

Although capital punishment no longer threatens the glassmakers, the best of Murano glass is sought after as enthusiastically to this day as it was at the time of the Medici's and the court of Louis XIV several centuries ago.

Glassmaking goes commercial

When the demand for glass containers began to exceed the ability to produce them in the traditional manual manner, glassmaking technology, especially in England, led to the split mold method of glass container production. This changed glassmaking from an emphasis on creating luxury goods for a restricted class of European society to mass production of bottles for wine and jars for drugs, demanded by the growing population in Europe and North America, starting in the 1700s. The addition of paper labels on bottles and glass vials to identify their contents and production origin gave birth to a thriving industry of commercial glass container production.

Then, in 1903, the growing popularity of glass bottles led to another significant and visionary packaging event, the invention of the first automatic bottle-making machine by Michael J. Owens, the founder of what is today Owens-Illinois, one of the largest international manufacturers of glass and glass products.

... and then came paper

Meanwhile, other architects of the Industrial Revolution were not sitting on their hands, content with materials that were available at their time. As glass production became mechanized, papermaking underwent parallel developments that made it a primary component of the rising importance of packaging. In fact, no raw material has done as much for the growth and importance of packaging as did the paper industry.

The name "paper" is derived from the Latin papyrus, a plant whose stems provided the basic raw material from which a paperlike material was produced thousands of years ago by the ancient Egyptians, Greeks and Romans.

Despite its name derivative, the very first paperlike material was actually made not from papyrus stems but by pulping fishnets and rags and, later, by a substance of plant fibers. A Chinese inventor by the name of Ts'ai Lun, who lived during the Han Dynasty, is most often credited with creating the first real paper from bamboo and mulberry fibers. This was encouraged by the Chinese Emperor Ho-di who had the vision of using paper to record events for future generations. The Chinese considered the ability to produce paper so important that – similar to the glassmakers of Murano – they kept the methods of manufacturing paper secret for over 500 years until the secret eventually leaked out to Japan.

Then, in 751 A.D., near Samarkand, an area that is today the Republic of Uzbekistan, in a battle at the Talas River, the victorious army of Ottoman Turks captured several Chinese papermakers and forced them to divulge the knowledge of their craft. From there, papermaking technology spread rapidly throughout the Middle Eastern region to Egypt and Africa and, around the 12th century, to the European continent.

Papermaking gains ground

When Arab legions invaded and occupied the Iberian peninsular, papermaking was introduced in Spain where, in 1151, the first European paper mill is believe to have been established. From there, the technology of papermaking expanded to France, Germany and England.

It was in England that the first "flexible packages" were produced, beginning with the production of paper bags produced from flax fibers and linen rags. This was followed in 1817, also in England, by the production of the first paper box.

In 1870, Robert Gair, an American paper bag maker, developed what turned out to be one of the truly visionary packaging inventions – the first automatically folded carton, a forerunner of the ubiquitous folding cartons. Folding cartons, in numerous varieties, soon became the most widely used form of packaging, dominating the packaging spectrum until the 1970s, when plastic packaging became popular with both manufacturers and consumers, challenging the paper industry's long-held predominance.

Meanwhile, corrugated paperboard was introduced in England in the 1850s, first in single layers and, ultimately, using two faces. Corrugated cartons using two-faced corrugated paperboard soon began to replace wooden barrels and shipping crates.

The production of corrugated paper even took on a political connotation. When growing paper production in North America caused English paper exports to the colonies to fall behind, England's Stamp Act (1765), which sought to make paper made in America less competitive, is believed to have become one of the issues that led to the American Revolution.

The resolute metals

Nothing quite so dramatic occurred in connection with the development of metal packaging. At least not for a while.

With all the attention on developing glass and paper into containers to hold various items, none were capable of preserving perishables, such as fish, meat and fruit. Metal containers had existed for thousands of years but most were made from silver and gold. The cost and difficulty of creating them made them unlikely candidates for the task of becoming a practical means of keeping food from deterioration.

It was not until the 13th and 14th century that tin-coated iron cans were first produced in Central Europe. But it required laborious steps to produce these and they were hygienically unreliable to preserve edible products. Meanwhile, like Murano glass and Chinese paper, the tin-coating process was kept such a well-guarded secret by its producers that few people had any knowledge of how to achieve it. Not until the 1600s, when the Duke of Saxony somehow acquired mastery of the technique, did tin-coating take hold in other countries, such as England and France.

Napoleon's contribution

As is so often the case, it is the urgency created by war that gives birth to visionary inventions. That, in fact, is what ultimately led to the creation of the tin can.

None other than General Napoleon Bonaparte, in search of being able to preserve food for his armies during his campaigns into Central Europe and Russia, promoted the idea of developing containers that could accomplish the preservation of food shipped over long distances. Napoleon needed containers that could prevent the deterioration of food for the lengthy periods of time that he needed to sustain his troops during his far-reaching campaigns.

In response to an offer of 12,000 francs by the French government to discover a method of preserving food for such lengthy periods, Nicholas Appert, a French chef and confectioner, experimented until he discovered that by applying heat to sealed glass bottles preserved the food within. Four months at sea with a variety of food thus prepared for the French navy convinced the French that they had, indeed, discovered the secret of sterilization.

From glass to metal

Across the Channel, the English went one step further. In 1810, Peter Durant developed a method for packaging meat, vegetables and fruit in airtight, tin-plated, wrought-iron cans for which King George III granted him a patent. Never mind that the cans required a chisel and hammer to open them and several days to produce, it was the visionary development of adding tin-plating to iron cans for preserving food that, although in greatly changed form, initiated and revolutionized the way we live and eat today.

The technology of preserving food through sterilization came to America when Thomas Kensett, an Englishman, emigrated to the United States and, in 1812, together with Ezra Dragget began canning oysters, meat, vegetables and fruit in glass containers, sealing them with corks. Finding glass too easily breakable and the corks too porous, Kensett, in 1825, obtained a patent for tin-plated cans from President James Monroe. He is therefore considered the founder of the canning industry in the United States.

Early canning was tedious. Cans were hand-filled through an opening at the top. After the food in the can was boiled, a disk was soldered over the hole to close the can

Nevertheless, canning in the late 1800s was a complicated and tedious procedure and could accommodate only a limited number of products. Metal had to be cut into the desired shapes, bent and soldered by hand. The tops and bottoms were then soldered unto the body, leaving a one-inch opening at the top that allowed small pieces of food or liquid to be fed through the hole. After boiling the contents, the opening was soldered shut.

Canning matures

Despite these initial difficulties, canning became a major industry that thrived with the continuous efforts by a growing number of entrepreneurs to find faster, cheaper and more reliable ways of canning perishable goods.

Replacing iron with tin-plated steel and using increasingly sophisticated manufacturing methods, can manufacturers in America soon found many applications beyond the daily staples of vegetables, fruit and juices. A growing variety of goods in metal cans, including beer, coffee, cookies, tobacco, toothpaste, pharmaceuticals and even cosmetics, soon gained wide distribution.

As the list of contributors to the canning industry gained in volume, Gail Borden, an early entrepreneur, is particularly noted as an innovator in the history of canned packaging. His experiments with canning, starting with meat, fruit and juices, eventually led to the canning of evaporated milk, an innovation that has significantly contributed to the health of children throughout the world and is still an important method of preserving milk in many countries.

The growth of the canning industry was aided by the progress made in printing technology. While the earliest efforts of identifying the can contents or decorating cans required soldering embossed labels to the cans or hand-painting the can surfaces – an artistically rewarding but tedious task – experiments with other, less time-consuming methods paralleled the maturing of can manufacture. Transfer printing was followed by lithographed paper labels and, eventually, by lithographic and flexographic printing on the metal can surfaces themselves.

At first, lithographic printing techniques during the late 19th century needed a great deal of experimentation. Using stones and metal as early carriers of the inks proved difficult as a transfer method to the hard surfaces of the metal cans. After much experimentation with colors and lacquers, the introduction of rubber plates and the invention of rotary presses simplified the adhesion of inks to metal can surfaces and allowed for a great variety of graphic applications. Soon, lithographic printing and, later, flexographic printing for decorative or promotional purposes were standard printing procedure on cans and other types of metal containers.

A new boy on the block: Aluminum

In the 1960s, aluminum cans appeared and quickly became the preferred material for cans over tin-plated steel cans, primarily because consumers preferred their lightness to steel cans. A dramatic improvement for aluminum canning came with the introduction of the draw-and-iron process, a technique which produced the type of two-piece cans used today by most beer and carbonated beverage companies.

The aluminum industry has extolled the advantages of aluminum over steel cans by claiming better chillability of beverages, improved stackability, resulting in better use of shelf space, and greater production speeds that reduce packaging costs for marketers and consumers. Aluminum cans became even more popular with the consumer when easy-open aluminum can lids were added to soft drink and beer cans, encouraging drinking straight from the can, thus eliminating the need for a cup or glass.

The biggest boost for aluminum cans came in 1967 when Coca-Cola and Pepsi-Cola switched to the two-piece aluminum cans, encouraging other soft drink marketers to follow their lead. Since then, aluminum can production, applied to many other products, has grown steadily, although it has been challenged by the inroads of PET (polyethylene-terephthalate) bottles for fruit juices, soft drinks and bottled waters because of the advantage of resealability of these plastic bottles.

More recently, the flexibility of aluminum has encouraged innovations such as the ability to shape and emboss the cans. Coca-Cola's shaped cans made a brief, although not entirely successful appearance. Heineken introduced a barrel-shaped beer can a few years ago and one major brewer is said to be testing the market for a can in the shape of a beer bottle.

In Japan, a country where the vending machine is a popular means of dispensing almost anything, shaped cans are making increasing inroads for everything from beer to hot coffee.

Aside from beverage containers, aluminum has its applications in the form of tinfoil for wrapping products such as quality candy or cheese or, laminated to the inside of paperboard folding cartons, to provide protection from moistness that paperboard alone cannot accomplish. Foil can also be found laminated to the surface of cartons when emotional appeal calls for the package to communicate luxury, such as on cosmetic packages, or simply as a promotional device of calling attention to the product.

The plastics revolution

As in every business, competition presents a constant challenge, and packaging is no exception. New materials, new manufacturing techniques, new products, new sales environments, new lifestyles, new consumer needs – all impact on packaging at every juncture.

No sooner had glass, paper and metal manufacturing techniques become more and more sophisticated, along comes a newcomer that makes trouble for all other packing materials but presents opportunities for packaging like no other material – plastics.

Plastics have initiated a true packaging revolution. Their translucency, formability and relatively low cost mean that plastics threaten to replace virtually all previously used packaging materials.

Although discovered in the 19th century, plastics did not reach their visionary status until the 1930s, when refinements by German chemical manufacturers created styrene foam for use in cups and food trays. From then on, the employment of plastic material for numerous functions began to gain popularity among manufacturers as well as among consumers.

Another war-time baby is born

The real boost for the proliferation of plastics came with the invention of acrylic in the 1930s. Its hardness and transparency led to its wide use in canopies for German, English and the United States aircraft. At the same time, the development and application of polyethylene and other plastic films for the preservation of food and the protection of electric cables during World War II led to the explosive growth of the plastics industry during the war. Their proliferation accelerated after the war ended and made them into one of the miracles of modern life.

A wide range of transparent films and other plastic materials for packaging, including cellophane, polyethylene (LDPE and HDPE), polypropylene (PP), polyvinyl chloride (PVC) and an ever-expanding range of derivatives have made plastics into the packaging darling that it is today. Many plastics soon became household trade names, such as Dacron, Mylar, Teflon, Plexiglas and Formica, to name just a few.

The phenomenon of plastic packaging for every type of product now covers the gamut from tiny blister packs for pharmaceutical products to foam padding for the shipment of sensitive instruments and heavy machinery.

Only recently, environmental concern has slowed the use of plastic materials for packaging. In some countries, such as Germany and France, strongly enforced laws govern the manufacture and disposal of plastic packaging, and experimentation with reuse or recyclability of plastic material is being extensively explored.

From peddling to merchandising

While the development of all these packaging materials – glass, paper, metal and plastics and combinations thereof – has been maturing, the gradual transition from the neighborhood grocery store to store chains that would reach across the breadth of the United States was taking place. Initiated by A&P and IGA, individual supermarkets began to spring up more and more, especially in urban areas, until the return to normal living conditions at the end of World War II gave birth to a plethora of new supermarkets and supermarket chains everywhere.

Within living memory, it is difficult to identify anything that has influenced the conduct of our daily lives more than the supermarket and its more recent successors, the hypermarket, as well as several other types of shopping facilities of hyper proportions.

Shopping at any of the mega-outlets has become the subject of everything from serious analyses at seminars attended by senior marketing executives to jokes about consumer buying habits.

The fact is that today it would be difficult to visualize our lives without these outlets. Whether you are in need of food, housewares, cosmetics, pharmaceuticals, hardware, electronics, toys, gifts, furniture – you name it – these stores are the lifeline of our modern lifestyles. Whether you consider it a necessity or entertainment, it's a good bet that supermarkets and mega-outlets will continue to shape much of our lives in the coming years.

And yes, even the much trumpeted shopping on the Internet, although a growing competitive method of acquiring goods, is not likely to supplant supermarkets and hyper-outlets for a long time to come, if ever. People may complain about having to go shopping but, to many of them, shopping is a welcome diversion from the daily routine of household chores and chauffeuring the kids to baseball and football games.

But what about packaging? With all these many shopping opportunities – supermarkets, hyper-outlets, specialty stores and the Internet – what role will packaging play in the future?

Packaging misused: A cure for every ill

As we stated before, supermarkets, in the sense that we refer to them today, did not develop seriously until after World War II when the needs of returning war veterans and their growing families created a new perspective on merchandising.

Small local merchandisers existed, of course, hundreds of years ago. But, strange as it may sound to us today, the origin of local stores was actually a reaction to the activities of overly ambitious traders in the nineteenth century who went from house to house peddling various merchandise, especially elixirs of questionable reputation and dependability.

This bottle of Sumlakia promised to solve disorders ranging from epilepsy to nervous disturbances, sleeplessness and hysterical conditions

Many of these so-called remedies came in fanciful packages. These packages proclaimed delivery from every type of malady you could think of. There was the package for Robertson's Infallible Worm-Destroying Lozenges, Mug-Wump Specific for the Cure and Prevention of Venereal Diseases and Clark Stanley's Snake Oil Liniment for eradicating rheumatism. Topping them all, there was Bonnore's Electro-Magnetic Bathing Fluid, whose package promised to cure everything from neuralgia, cholera, rheumatism, paralysis, hip disease, measles, female complaints, necrosis, chronic abscesses, mercurial eruptions, epilepsy and scarlet fever. Wow!

No wonder that packaged products of those days earned the reputation of containing untrustworthy products. It's to the credit of a few more responsible manufacturers that some medical products, such as Carter's Little Liver Pills, Smith Brothers' Cough Drops, and Johnson's Baby Powder,

In contrast to the exaggerated claims of early medical packages, Johnson's Toilet and Baby Powder simply described its usage "For Toilet and Nursery"

survived the unsavory reputation of those early patent medicines and are still available today.

Gaining confidence in packaging

So, to gain the confidence of consumers at that time, some merchants opened small local stores where they sold food, tobacco, hardware and a variety of other merchandise over the counter to neighbors who lived in walking distances of their stores.

These entrepreneurs took pride in tending to the needs of local neighborhoods. Their services were very personal. They recognized most of their customers by name and their customers knew and trusted them. Their products were often wrapped by the storeowners or their clerks or fished out of barrels that cluttered every inch of the stores.

Early grocery stores created a jumble of barrels, cans and bottles covering every inch of their limited space, some merchandise even hanging from ceilings

But, as early as the 1870s, a few items, such as baking soda, biscuits and soap, started to be prepackaged by the producers of the products.

Gradually, some manufacturers and suppliers of these products began to realize that packaged goods not only protected the products from deterioration and damage but were a way of calling attention to them, so that the people who bought them were likely to buy them again by remembering their brand names and their package appearance. Thus, the idea of identifying a product by brand was born and packaging became its main carrier.

Among the early pioneers, some born in late 19th and early 20th century, who recognized the effectiveness and took advantage of this new method of merchandising were brands such as Quaker Oats, Cream of Wheat, Hershey, Coca-Cola, Morton Salt, Salada Tea, Underwood Deviled Ham and Ivory Soap. These brands, all with humble beginnings, are among those which, in time, grew into mega-brands and category leaders because they were among the first to understand the vision of branding.

Visionary packaging comes to America

But packaging on a large scale and as an indispensable marketing tool did not emerge until the early part of the 20th century. All this time, more and more products became available and packaging graphics became an economical and effective way of promoting what was inside the packages. As small stores, once owned and operated by individual entrepreneurs, expanded into managed chains of retail stores and eventually into mega-stores with hundreds of outlets, packaging became an essential component of the shopping experience.

One of the first major merchandisers was the Great American Tea Company, founded in 1859 as a mail order business by tea and spice merchants George Huntington Hartford and George Gilman. But it was not until 1861 that their first store opened in New York City. This was quickly followed by the opening of many more stores so that, by 1867, it represented the first major grocery chain in the U.S. In 1870, the company was renamed The Great Atlantic and Pacific Tea Company (A&P), forecasting its aspiration of marketing coast to coast.

The foresight of the people who managed the early A&P stores is evidenced by their vision of introducing the first private brand labels available at any grocery store. By introducing Eight O'Clock Coffee and Our Own Tea, A&P acknowledged their early recognition of the power of branding to attract shoppers into their stores. Eight O'Clock Coffee, in several varieties, is still available today.

A&P's introduction of an own brand, Eight O'Clock Coffee, was a visionary first in the grocery business

Another prominent retail development was the creation, in 1926, of the Independent Grocers Association (IGA). This association of a number of independent grocers provided individual storeowners with the advantage of purchasing merchandise as a cooperative in large quantities and at lower wholesale prices. In this way, they were able to market more profitably vis-à-vis individual grocers. At the end of 1925, more than 150 IGA stores had affiliated in this manner. By 1930, the number of IGA stores had swelled to over 5000 and, like A&P, quickly jumped on the bandwagon of marketing their own lower priced private labels in competition with national brands.

Today, virtually every supermarket and hypermarket chain offers their own labels and has made these a major source of income, The impact of private labels has affected the lifestyles and shopping habits of everyone in the civilized world.

Chapter 3
The commercial power of packaging

When, as early as the mid-1800s, many storeowners began to appreciate the power of the package, they quickly understood that handling and offering merchandise in prepackaged form was the most efficient way of merchandising and maintaining their stores. With multiple stores and store chains, the control of distribution, shipping, storing, displaying and selling a large number of products of every description, packaging became an indispensable medium without which it was no longer possible to operate effectively.

It did not take long for more and more manufacturers of products to understand the opportunity of promoting their products by packaging them in colorful containers that provided a platform for recognition of their brands, promoting the benefits of their products and catching the attention of the consumer at the point of sale.

Many of the companies that have a leadership position in their categories today were among the early pioneers who had the vision of providing effective packaging that recognized and addressed the needs of consumers.

Brands such as Arm & Hammer, Mennen, Aunt Jemima, Gillette, Budweiser, Heinz, Campbell's, Ralston and Nestlé, were among the packaged goods that were born around the turn of the century and have survived to this day.

There are many fascinating examples of early pioneers of packaging concepts that led to the way we all shop today. Some of these package developments are remarkable for their visionary foresight.

From soup to art

In 1869, fruit merchant Joseph Campbell and icebox manufacturer Abram Anderson formed a partnership that offered a wide range of canned products, including several vegetables, jellies, condiments, mincemeat and their most successful product, Beefsteak Tomato. Already wise to the effectiveness of promoting the products on their labels, the cans proudly announced that the products were made "by the rich perfection of our goods".

Before Tomato Soup came "The Celebrated Beefsteak Tomato" by Campbell

Photo courtesy of Campbell Soup Company

When Abram Anderson left the company, it was renamed The Joseph Campbell Preserve Company. But it was only when Dr. John T. Dorrance, a 24-year-old relative of Joseph Campbell, joined the company that things really began to happen. Dr. Dorrance had just returned from studying in Germany where he fell in love with the soups that were highly popular there. The introduction of Campbell's ready-to-serve Beefsteak Tomato Soup was soon followed by a condensed version of Tomato Soup in a $10\frac{1}{2}$ ounce can with an informative, but visually undistinguished label design.

The label of the first condensed version of Campbell's Tomato Soup still referred to the Beefsteak Tomato

Photo courtesy of Campbell Soup Company

It was a football game that led to one of most visionary package design concepts ever. When, in 1898, Dr, Dorrance attended a Cornell-Penn University football game, he was so taken by new red and white uniforms of the Cornell team that he adopted the colors to create the now famous red and white Campbell's Soup label. Even the wooden cases in which, at that time, Campbell's Soups were shipped to a growing number of grocers who sold their products, carried the red and white label design.

One of the first to use color for brand identification, Campbell's red and white labels have become an icon for good soup and led to one artist's worldwide fame

Photo courtesy of Campbell Soup Company

To this day, the red and white Campbell's labels, although having undergone numerous design modifications over the years, continue to have such powerful equity that Campbell's Soup virtually owns the soup market in the United States.

And – think of it – how many commercial labels can you name that have the distinction of being displayed in museums all over the world as the result of becoming a major, recurring subject for the paintings by a well-know modern artist?

Pass the ketchup

At about the same time, in the 1876, not far away from where Campbell's Soups were being produced, another visionary packaging event was taking place. In Pittsburgh, Pennsylvania, a young lad by the name of Henry Heinz founded the H. J. Heinz Company, producing 57 varieties of products, including what would eventually become the world-famous Heinz Ketchup.

The original Heinz Ketchup did not anticipate the ultimate shape of today's well-known Heinz Ketchup bottle. The original glass bottle was shaped like a keystone, relating to the keystone symbol of the state of Pennsylvania, where the product was produced. However, what did

anticipate the future bottle structure was a long narrow neck that was created for two reasons, to reduce the contact with air that tended to darken the sauce and – believe it or not – to make it easier to pour! It was a visionary concept, even when we have become all too familiar with the need to vigorously, but lovingly, slap the bottom of the bottle to dislodge the ketchup.

At the turn of the century, Heinz Ketchup glass bottles already featured the distinctive long neck that has since become a Heinz icon

Photo courtesy of H.J. Heinz Company

After several bottle and label changes during the following years, a 12 oz. octagon-style glass bottle with a screw top was introduced around the turn of the century. It retained a long, sloping neck that, alas, resisted dislodging the product as much as did its predecessors.

But, like it not, the styling of the Heinz Ketchup bottle has remained the trademark of the brand. Other bottle configurations for Heinz Ketchup have been attempted from time to time, and several other versions are currently available, including an upside-down plastic bottle that, while retaining equity through the keystone-shaped label, exchanges familiarity for storage and dispensing advantages. Still, none of the modern shapes have replaced the popularity of the octagonal glass bottle. Despite all its difficulties with pouring, the vision of H. J. Heinz to stick tenaciously with the long-necked bottle is, in part, responsible for making Heinz Ketchup the undisputed favorite condiment among all generations and around the world.

The plastic Heinz Ketchup bottle retains two Heinz Ketchup icons: a long neck and the keystone-shaped label

Except for the keystone label, Heinz Ketchup recently broke with tradition with an upside-down plastic bottle that exchanges familiarity for storage and dispensing advantages

Green and white = no headaches

On the other side of the Atlantic Ocean, the Industrial Revolution produced its own visionaries of new products and new packaging, many of which have grown into well-known, internationally distributed commodities.

One of these, Bayer Aspirin®, was one of the first to have the vision of using packaging not just to facilitate product distribution and protect the product, but as a means of legally protecting the brand's trade dress. A derivative of dyes used for coloring silks and cashmeres, Phenacetin Bayer, the first Bayer drug, is credited for having saved the lives of countless people during the 1889 flu epidemic that raged in many countries in the northern hemisphere. Small dosage packets were distributed to druggists in cartons that anticipated potential counterfeits by stating, in three languages:

Resale only allowed in this original packaging.

Strangely, the labels also warned that "The Resale and Importation to the United States of America, Canada and Italy are prohibited." Whatever the reason for this restriction, most likely patent-related, the notice indicates that packaging already played an important role in branding by dealing with product counterfeiting in the non-regulated drug market of the 1890s.

Aspirin, a name derived from the combination of acetylsalicylic acid and spiric acid, first appeared around 1899 in powdered form in a 250g glass bottle with a cork closure. Available in tablet form around the turn of the century, Bayer Aspirin packaging went through a series of structural packaging modes, progressing from tubes to metal containers to paperboard cartons.

As Bayer Aspirin grew in popularity it has become the most widely used analgesic around the world.

But, as with all brands that are handled by many hands in many countries with different languages and local customs, graphic liberties eventually fractionated the brand look to a point where proprietary brand identity, as well as legal ownership of Bayer Aspirin was in jeopardy. Our company was asked to review the packaging graphics and to recommend a design system that would correlate the graphics of Bayer Aspirin packages all over the world and regain the brand's visual equity. Now, except for the U.S., Canada and a few other countries, a comprehensive identity system reconciles a variety of languages, typographic styles and local characteristics, while carefully retaining the key recognition elements of Bayer Aspirin's international look – the bright green and white colors, black product logo and the Bayer corporate logo.

This visionary step by Bayer to strengthen worldwide visual recognition through a coordinated package design program was therefore

not just a surface enhancement, but strengthened its recognition and reputation for efficacy among druggists and consumers and fortified Bayer Aspirin's constant battle against look-alike packaging, especially in the private label sector.

Around the turn of the century, Bayer Aspirin began offering its medication in tablet form in a glass bottle with a cork closure

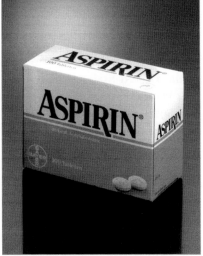

Except for the US and a few other countries, the green and white trade dress now identifies Bayer Aspirin in pharmacies all over the world

Photos courtesy of Bayer AG

"Persil bleibt Persil"

Fritz Henkel, another European entrepreneur, was responsible for products that have dominated the detergent category in Europe to this day and for which packaging has been a critical component.

Starting with the introduction in 1876 of "Henkel's Wasch und Bleich Soda", a universal detergent with a bleaching agent for hand-washed laundry, superior to other washing agents available at that time, the product was distributed in bulk paper bags with copy that promised to achieve brilliantly white laundry without harming hands or laundry.

The success of this product quickly led to its replacement by an even more effective detergent, introduced in 1907 in paperboard cartons under the brand name of Persil, a derivative of two key product ingredients, perborat and silicon. The carton copy promised effortless results without the need

Henkel's early paperboard packages promoted the convenience of combining washing detergent and bleach

Photo courtesy of Henkel GkaA

of bleaching or scrubbing and being harmless even if misused – such were the concerns in the days of scrubbing laundry on washboards, prior to the invention of mechanical washing machines.

More importantly, the visionary nature of that early carton was its strong use of green, white and red colors that have continued to identify Persil packaging to this day. The choice of green was presumably a reference to the previously traditional need for bleaching laundry by spreading it out on lawns. White communicated the whiteness of the clothes washed with Persil. Red called attention to the words "Washing Agent" and, later, evolved into the color for the brand name on all Persil packages, the exception being the packages of the environmentally oriented phosphate-free "Green Persil", introduced in 1985.

Persil, now available in many countries in Europe, the Middle East and the Far East and in a variety of forms and packaging, has tried hard to maintain its brand equity. But, as with all successful products, it did not take long for numerous counterfeits with similar sounding brand names and similar looking packaging graphics to appear on the market.

Recent packages boldly emphasize the Persil brand name as an icon that identifies Henkel's category leadership in Europe and around much of the world

Photo courtesy Henkel GKaA

Henkel defends its product with the slogan "Persil bleibt Persil" – translating literally "Persil remains Persil" – inferring that only detergents in the well-known green, white and red packages are the genuine Persil,

When, in the 1930s, cheaper detergents came onto the market *without* packaging, claiming to be loose Persil, Henkel considered this serious enough to warn consumers with a poster that appeared on advertising kiosks throughout Germany:

> Warnung! *Es gibt kein loses Persil!* Persil ist nur in der grün-weiß-roten Packung – mit dem Aufdruck Henkel im roten Felde – zu haben, niemals lose!

Translated into English, the poster is a testimonial to the visionary brand identity and packaging:

> Warning! *There is no such thing as loose Persil!* Persil is available only in the green-white-red package – with the brand name Henkel in the red field – never loose!

Available only in the green and white package with the brand name Henkel! What greater evidence could testify to the critical importance of visionary packaging to preserve the recognition and authenticity of a brand?

A changing world

Since the days of the peddlers, packaging has unquestionably grown substantially in importance to product marketers and in popularity with consumers. Still, with the exception of a few enterprises that had risen to the proportions of chain stores , such as A&P and IGA, the sale of packaged products remained mainly the trading method of neighborhood stores throughout the 1920s and 1930s.

Once again, it took a war to dramatically change the direction of our lifestyles.

As World War II veterans returned home and acquired homes to accommodate their growing families, the costs of new homes and the arrival of children sparked a rash of do-it-yourself activities that initiated an explosion of packaged products in home centers and hardware stores.

Homeowners acquired more and more tools to help accomplish tasks previously handled by plumbers, painters and electricians. Growing and more affluent families generated the demand for more and more houseware. Fixing and servicing cars, once the sole domain of car dealers and service stations, became manageable by anyone handy enough to perform these tasks with the help of thousands of do-it-yourself products available at automobile after-markets. Meals, both prepared and home-cooked, were made increasingly easier and more varied for the growing community of dual-income families through the availability of an ever greater variety of canned, bottled and boxed products.

With all this hectic postwar activity, packaging quickly took center stage. Various products and product categories required new packaging forms, materials and visual concepts that tried to address different consumer audiences and consumer needs.

The ranks of professional specialists in package engineering, brand identity, package design and consumer research, sought after by manufacturers, marketers and retailers, swelled during the 1960s and thereafter. The introduction of computers in the 1980s, providing greater versatility and speed to the package conceptualization and production process, turned an entirely new page in the history of package development.

As lifestyles change

What has affected packaging most – and continues to do so – is that the lifestyles of modern families have undergone major upheavals in the years following the end of the war. Mom and dad became dual-income earners to afford all the things they want to own and pay for the schooling and the needs of their growing family. No less important are the changes in the distribution of goods among divorced or separated family members.

Most significantly, without any doubt, is the invasion of cellphones and computers into our private lives, making those of us long accustomed to domesticity accessible at any time and at any place. This has turned our lives upside down, whether we accept these changes enthusiastically or prefer to hold onto a more traditional lifestyle.

Maggie Jackson, in her book *What's Happening at Home – Balancing Work, Life, and Refuge in the Informative Age* puts it this way:

> Our way of working, the way we think of marriage, how we communicate with each other have all been transformed. We are coming out of an era of human society that was much more focused on boundaries. More and more people are eating individually in their houses. Many are not gathering anymore around the dinner table. It's not even that mother is leaving dinner for the kids. The children are actually making their own food that consists of snack food, pop corn, or frozen this or that, something they can grab, walk around with or eat in their room.

If Ms. Jackson, herself a mobile mother, is onto something, it's not hard to see how such lifestyles have affected, and will continue to affect, packaging of goods that cater to our mobile needs on several age levels. There is no longer just mom making the decisions. As the mobile family grows or becomes separated, packaging has to reach out to a lot of different constituents – adults, teens and children.

Mobility and affluence affect how we live

As Ms. Jackson sees it:

> The other thing regarding mobility is the idea of the summer house. There are now six million second homes in the United States. People are dividing time between homes, leading more and more to some kind of divided lifestyles. They need two of everything – two sets of clothing, two computers, two refrigerators, double the quantity of food and beverages and so on.

We have already noticed this in the stores. Prepared meals and snack foods are becoming more numerous and available in greater varieties in supermarkets and mega-outlets. The bundling of products, from beverages to batteries, from apparel to stationery supplies, has become the norm – the myriad of brands in virtually every category is mind-boggling and private labels of supermarket and mass-merchandising chains, from food to electronics, are proliferating.

As mobility increases, and as three and four car families are no longer exceptional, packaging will play the role of making take-out meals and fast-food dining more and more an accessible standard.

The meteoric rise of private labels

In the midst of all this lifestyle turmoil is the meteoric rise of the private label. Taking the cue from A&P and IGA, private labels have gone from the perception of low-priced competitors of corporate brands to a position of brand leadership.

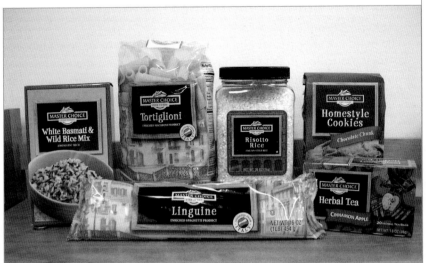

Private labels by supermarket and hypermarkets have not only become more and more numerous, but rival, and often surpass, the sophistication of manufacturer's brand packaging. Master Choice at A&P is a good example

"What started in the 1970s," says Brian Sharoff, president of Private Label Manufacturers Association, "has now reached the point where retailers no longer say that they sell private label or corporate brands or retailer brands, or exclusive brands. The consumer no longer sees the difference between proprietary and private brands. Price points and quality are no longer as far apart. If you walk through stores – whether it's H-E-B in Texas or Monoprix in Paris – you see packaging that is competitive with what the corporate brand used to think was their own domain, and is not their domain any more."

As many grocery retailers are thinking in terms of expanding into new categories such as housewares, wines and even computers, Sharoff opines:

The single most important challenge that the manufacturer of private label has is to provide the retailer with the kind of packaging that meets consumer expectations. In the next decade, the only element to differentiate private labels from brands will be the packaging, and he who comes up with the best packaging will have the advantage.

Making a point of difference

The trend towards greater emphasis on private labels, or "own labels", as they are called in the UK, has made an impact on the sale of consumer package good brands whose advantage over store brands has diminished in recent years. This is not only the result of changing consumer attitudes, but a change in store management who sees packaging as a tool to convey the personality of the store.

"What is changing," explains Perry Seelert, VP, product strategy at Daymon Worldwide, "is that retailers are beginning to realize that it is not just procurement and buying efficiencies and back door efficiencies, but that it is really the front end, the consumer, who is going to be important in the future. It's not enough to have private labels to enhance margin. Consumers want to perceive a point of difference in the store brand. That's what is driving proprietary branding today…Europe has a philosophy of driving own brands. Dutch retailer Ahold, German retailer Tengelmann, French retailer Carrefour, all have seen own brands to be successful in their homelands, and they see it not only just as margin enhancement but as a means of store differentiation."

Building consumer equity

Although, as Perry Seelert indicates, the use of packaging to focus on the personality difference of one store chain vis-à-vis another is a strategy followed by many major chains, their methodology of attracting customers varies greatly.

Tesco, one of the leading store chains in the UK, acknowledges a three-tiered own label system that tries to squeeze manufacturers' brands between their lower priced and higher quality pricing spectrum.

Most private brands in the U.S. seem to prefer a two-tiered private brand label approach. A&P, for instance, offers an economy brand,

America's Choice, together with Master Choice, their higher quality offering. America's Choice includes multiple product categories from household products to food, pet and baby items, while the role of Master Choice is to be a more upscale, specialty food line.

Whatever the system, the growing emphasis on private label branding clearly identifies the critical role that visionary packaging will play in future marketing strategies for major supermarket and mass-merchandising chains.

the commercial power of packaging

section two

the consumer
mindset

Chapter 4
Package design and the consumer

Never judge a book by its cover.

How often have you heard this? But it's true, judging a book by its cover is not the best way to assess the book's contents. But we do it anyway.

Nor should a person's capabilities be judged by the way he or she is dressed. But, more often than not, that's exactly what we do. It's a habit we've grown up with. We are visually programmed to do so and we don't easily change idiosyncrasies or biases.

Meet John Smith wearing a plaid shirt and jeans and you will see him in your mind as a relaxed individual. Meet John Smith a few days later at a business meeting, dressed in a black suit, white shirt and silk tie and you get a totally different impression of him, even though it is the same person.

What applies to judging books and people applies equally to packaging. We judge what's inside the package by the design of the package. It's an automatic emotional reaction. We select products by the perception of what we see and read on the package. Is it the brand I want? Is it the flavor I want? Does it look like the quality I want? Is it right for me?

It's a fact. More often than we may like to admit, it's the design of the package that drives our purchase decisions. The package design bends our mind. In an environment where the onslaught of visual and audible messages engulfs us every minute of the day, we are an easy target.

How does this work? What role, if any, does package design really play in this mind game?

The silent mind-bender

Almost every time you open a newspaper or watch TV, there is some-thing about a religious or political group of people whose lifestyles have been influenced by powerful propaganda. Their minds are being conditioned to envelop new, provocative and often unconventional ideologies they may not have previously held. The results can be either beneficial or destructive. It's the ability of an individual, or a group of individuals, to bending people's minds by conveying ideas that influ-ence them to react in a proactive way. Do it often enough and passion-ately enough and you will have people eating out of your hand.

Just think of Moses, Jesus, Mohammed, Caesar, Hitler, Lenin, Gandhi. They were all powerful mind-benders, for better or for worse.

Package design is a mind-bender, although, thankfully, in a less con-troversial context. Unlike the personalities who have molded history by bending the minds of their followers through provocative speeches, packages have no audible voices, at least not yet. They don't gesticu-late. They don't have a body language. They don't attract large crowds and don't require sacrifices.

Yet, by being informative, provocative and seductive, package design can produce product personalities that communicate product attributes in ways that influence consumers to select Brand X over Brand Y. Just like a skilled salesperson who, through words and body language, explains to you the benefits of one product over another, the package that best communicates the advantages of one brand over competitive brands will seduce you to select that brand.

Package design is truly a silent mind-bender.

It's really quite amazing

Being able to influence the human mind, unaided by any direct human contact or voice, is a feat that has few parallels. But that is exactly what the design of a package is capable of doing.

It's also what those who want to sell their products have in mind for it to do.

When you think about it, this is a rather fascinating, almost myste-rious part of our daily lives. Hardly a day passes when you do not pick up a product in a package. As we said previously, you usually don't study the package or analyze its design. Unless you are looking for specific information about the product, you give the package design no particular thought. After all, it's the product you are buying. So you pay little attention to the package – or do you?

Just think about it. When you buy a purse for yourself, the store clerk will put it into a bag for you to carry it home. The bag may have the name of the store printed on it, but this is usually immaterial since the bag is just a convenient means of transportation.

But if you buy the purse as a birthday gift for a friend, would you hand it to her in that bag? Not very likely. Think Tiffany! You will ask the store to put it into a nice box with a ribbon around it. All of a sudden, the product has taken on a different personality, the personality of a valuable gift. What's more, this simple change of packaging has raised the perception of the giver in the receiver's mind.

It's a fact. Every time you acquire a product, whether an inexpensive household product or a luxurious gift, it's the package that gives the product a personality. It is that personality that determines your perception of the product inside the package.

Forming a personality

How does a package form a product personality?

Packaging serves many functions. Its design has the ability to shape your subconscious vision of the purpose, quality and benefit of the product. With the assault by television, newspaper ads, catalogs, brochures, websites, road signs, the Internet and thousands of products in mega-stores, supermarkets, department stores, boutiques, and who knows where else, all competing for our attention, packaging takes on the role of the ultimate decision maker.

Conveying the brand personality through the package must therefore be the prime objective of marketers of retail products and their brand and package design agencies.

It's as if you were meeting a person. As we said before, you tend to form an opinion about a person by the way they act and by the way they are dressed. It's the same with packaging. The design of the package creates a brand personality that instinctively forms your impression of the product inside the package.

The options are limitless. The shape of the package, the size of package, the graphics, the color, the text – all can be combined in innumerable ways to tell you something about the product and shape the product's personality in your mind.

It's an opportunity and a challenge for the marketer and the designer to be visionary, to create a brand and product personality

In the shopper's mind the package transmits an image of the product

that will create a favorable perception about the product in the consumer's mind.

Let's look at some of the opportunities.

The image-creating package

Packaging can convey different images to different people. It can communicate whether a product is meant for a male or female consumer. The colors and the shape of the package will appeal to feelings about self-identity that differ between most men and woman.

The container can convey the image of the product's worth. Take Chanel No. 5. Would you market it in a plastic bottle? Don't even try it! Or would you package common nails in a gold foil carton? Not likely either.

Such packages simply would not match your perception of the personality of these products.

Put an expensive necklace into the blue setup box from Tiffany. Then put the same necklace into a carton from Sears. Which of these will convey the true value of that necklace? When you see the Tiffany box you know that it holds something precious without even opening the box. The Sears package sends quite a different message. The package alone can convey a totally different personality for the same contents.

Tiffany's blue box communicates a valuable gift even before opening the package

Who would not be able to recognize these products by the package shapes alone?

The container shape can aid in identifying a brand. Could anyone fail to recognize Coca-Cola just by the shape of its bottle? Or Odol Mouthwash? Or Listerine?

Perfume bottles cater to women's desires. Talk about mind-bending. The diamond-shaped stopper on a Chanel perfume bottle implies luxury and value. The suggestive lovebirds on Nina Ricci's L'Air du Temp bottle hint at love.

The lovebirds on L'Air du Temps perfume bottles evoke a feeling of romantic affection and surrender

The functional package

The package can also be an integral component of using the product by providing convenience. Try selling a woman oral contraceptives in a package without the day reminder.

Or packing the lunch box for your kids with one of those little beverage containers without the straw attached to it.

Can you imagine life today without aerosol containers? The ease of dispensing liquids, whether paint, shaving cream, insect killer or room deodorizer, the simple action of pushing a button has become a modern convenience that we all take for granted today.

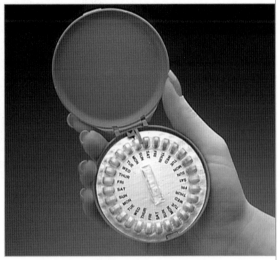

Beyond identifying brand and product, some packages fulfill functional tasks. This package for contraceptives functions as a daily reminder

But all the glitters is not gold. Despite the great popularity of this medium, the damage that aerosol propellants may do to the ozone layers around our planet earth gives many consumers second thoughts about its benefit. The next ten years may see a drive, by government sources or consumer initiatives, to limit the use of aerosols as a spray medium for numerous liquid products and lead to the development of new, less damaging propellants and packages to accommodate them.

The communicative package

Then there are names. Brand names. Brands that have been used smartly and consistently can communicate product perceptions of incalculable value. Just think of the brand names like Sony, Johnson & Johnson, Nestlé, 3M, Lipton, Krups or Bayer. These names have been around for many years and, through good brand management and good packaging, have succeeded in building and maintaining unmatched

reputations. What consumer would not expect consistently high quality from products in packages with these brand names?

Al Riess and Jack Trout, in their well-known book *Positioning*, offer this great example of brand name effectiveness: "Kraft has been successful in cheese. Now name all other cheese brands you know."

Logos do more than merely identify brands – properly managed, they are an investment in communicating images of quality and functionality

No-nonsense boldness communicates the perception of hard-working effectiveness. That's why detergents, such as Tide, Cheer or Persil, have labels that scream their names in bold lettering and multi-colored hues to grab the consumer's attention.

Kraft Cheese, a brand that almost solely dominates the cheese category in the U.S.

Visuals on packages are a must for some products, a promotion for others. Many packaged products, from dinnerware to hair coloring, are often selected by the picture on the package. Many foods and food ingredients benefit from pictures that show what the consumer should expect the finished product to look like. It may flatter some products beyond reality. But wouldn't you wear your best suit or dress when going to an interview for a job, even if this were not your everyday manner of dressing? Packaging is like dressing, it is presenting the product to the shopper as attractively as possible.

There are, however, instances when the need for conveying critical information about the product does not lend itself to pictorial representation. Pharmaceutical products, for example, fall into this category. The need for reversing an unpleasant physical condition that is the criterion of most pharmaceuticals is often difficult, if not detrimental, to portray.

Emphasis on brand identification may, in some instances, be more critical for some products than any kind of promotional presentation. In many cases, the balance between emphasis on brand identification or emphasis on product presentation may require difficult decisions.

A good example of this is Campbell's Soups. No one will question that the red and white labels of Campbell's Soups present one of the prime examples of superb brand identity. But, for many years, loyal Campbell's Soup enthusiasts used to complain about the difficulty of finding their favorite soup among the sea of red and white.

As early as the 1970s, we showed Campbell's Soup management the idea of adding an illustration of a plate of soup onto the label, in order to make the selection process easier for the shopper.

A logical step? Not so, thought Campbell's management at the time. They were convinced that the strength of the red and white label design of Campbell's Soup labels was all that was needed to attract shoppers. Food illustrations, they felt, would tend to compete with their brand identification, minimize their trade dress and potentially jeopardize their domination in the category.

No real surprise. When a brand that has been as successful as Campbell's Soup has been marketed in virtually the same red and white dress since around the turn of the century, it is a difficult decision to experiment with a novel approach.

It took the introduction of Campbell's Chunky Soup to demonstrate that soup illustrations on their packages were a visual cue that soup shoppers were looking for. Most Campbell's Soups have since joined

*Campbell's Chunky Soup
labels were the first to
respond to consumer needs
for visually identifying
the soups*

their Chunky Soup cousins and are now showing product presentations on most of their labels.

Many kinds of products need to convey product appearance, by being seen through carton windows, blister packs, film wraps or other types of packaging that reveal part of or the entire product. Toys, tools, electronics and household products are among those that consumers usually want to see before they purchase them without having to open the packages.

The sensuous package

What will probably come to your mind when you think about products that provide a sensuous reaction? It's good bet that cosmetics and fragrances will do so.

Many cosmetics and fragrance packages are designed to convey a sensuous reaction, such as we associate with beauty, desire, love, seductiveness and luxury. All have one common objective – to create the perception of "the good life".

But sensory reaction is not limited to beauty aids. All packages can create some sort of sensory expectation as to the product inside the package, whether it is a stick of chewing gum or a high-priced piece of jewelry.

Even the lowly shopping bag can create a sensory response. Shopping bags can communicate the perception of what you bought and where you bought it. Beyond simply being a means of transporting objects from a store to your home, shopping bags provide the opportunity to convey the special character of a store or of a brand.

Observe a lady walking in front of you carrying a Gucci bag. You wonder, is what she carries a present for herself, her husband or a special friend? Whichever it is, you *know* it's a precious item inside. Watch another person walking by with a bag from Cosco or Target and you *know* that you have met a bargain hunter.

Carry a Gucci bag and people will see you as a discriminating shopper

Carry a Target bag and you are seen as a bargain hunter

The package colors

Probably no other part of a package can influence your perception as much as the package colors.

Dark colors, such as most coffee packages, communicate the richness and quality of a warm beverage. Light colors, such as used on many dairy packages, are meant to convey health and purity. Yellow, red and orange communicate fun. The riot of primary colors on most toy packages imparts the perception of activity and excitement for children.

Package colors can also be functional tools. Buy an aerosol spray paint and you will select it by the color of the cap. Food label colors help to differentiate product varieties. Some will do so through association with the color of the product. Others will simply be a means of separating one product from another.

Package colors can also be used as a powerful mode of brand identification.

What Campbell's Soups' original red and white labels may have lacked in easy product differentiation, they unquestionably gained through instant brand recognition. So does Kodak who built an empire on their yellow packages. To escape the doldrums of red cola packages, Pepsi, in a bold departure from category tradition, switched to blue cans and labels and multi-packs to achieve a trade dress that now clearly differentiates Pepsi from other cola brands.

Following in the footsteps of Kodak, Bayer, and other companies that use package colors for branding, Leapfrog's green packages stand out in toy stores

Designs by Damore Johann, San Francisco, CA

Using color as a branding device is not necessarily a monopoly of major brands.

Take Leapfrog Enterprises, a medium-sized educational products company, located in California. To deal with the color oversaturation in toy stores, where most of their products are sold, they selected the color green for rebranding their entire marketing arsenal, including packaging, point-of-sale displays, brochures and print advertising.

"Selecting green as Leapfrog's identifying color was a big risk," explains Rita Damore, president of Damore Johann Design, Leapfrog's design firm, in an article in *BrandPackaging*. "But it was also such an obvious opportunity to differentiate Leapfrog from other educational products companies. No one else was using it. By selecting green as an identity icon, Leapfrog did in the toy industry what Barbie did with pink."

Testifying to the success of this program, Leapfrog educational toys are now available in 25 countries and the product arsenal has been extended to included electronic learning aids for youngsters in grades three through twelve.

Can there be any doubt as to the significance of package structures, colors, graphics, pictures and text to transfer a company's products from the shelf into the shopping basket?

The central role of package design in a changing market

With so many alternatives and potential solutions, and a constantly changing retail market, you can't help but wonder what packaging will be like five or ten years down the road.

Undoubtedly, mass-merchandising outlets will continue to flex their muscles and the Internet will continue to reshape much of our manner of trading. The puzzle is, in what way will our methods of selling and purchasing change the way we present our merchandise in the next few years?

Before we try to predict what packaging may be like in the next five or ten years, let's take a look at what package design has accomplished during the *past* few decades.

Marketers, it must be said, have generally been a conservative lot. There are few risk-takers among them. With a few exceptions, brand and product managers prefer to play it safe, keeping things neat and package design conservative.

You get the idea. Don't rock the boat. If it ain't broke, don't fix it.

Dramatic inroads are not a characteristic of the packaging trade, especially not among major corporations. Virtually all cereals come in the same folding cartons that have been used for decades. With a few exceptions, milk continues to use the same stock cartons and plastic containers. Round cans still predominate on supermarket grocery shelves even though they are not space-friendly when shelf space is at a premium.

OTC drugs remain partial to traditional stock bottles and hard-to-open blister packs. One of the favorite packaging media for hardware and some household products remains the clamshell, even though it is probably the single most opening-resistant form of packaging ever devised.

Unfortunately, exploring visionary packaging, based on consumer needs, rather than on manufacturing and production priorities, is a territory left to smaller companies, seeking a niche to explore for their products.

You can rationalize anything

Of course, you can rationalize all this. There are development costs, investments for new equipment, materials' accessibility, production economies and the expense of production line modifications for new and untried packaging systems. All these are certainly legitimate concerns to support conservative package design.

But the 21st century will have no patience for rationalization.

Look at the electronics industry. Where would it be today if it had let economy and production efficiencies dictate its product development? Consider the tremendous risk taken by Apple Computer Company when it decided, in the midst of the fiercely competitive PC market, to launch the iMac on a single platform and in five fashion colors.

Despite its current problems, the reason for the meteoric rise of the leaders in the computer industry, such as Dell, Apple and Microsoft, is that they look to the future without blinking. So quickly are new products explored, created, manufactured and launched, that it has become a cliché to say that the day you buy the latest model, it's already outdated.

Regardless, the electronic market thrives. Obsolescence is not an obstacle for it. On the contrary, it becomes an incentive for every new idea. There are new products virtually every day and there are buyers for them every day.

It's a take-charge attitude. It's an industry that believes that not taking risks is to close the window to the future.

Chapter 5
Consumer attitudes and concerns

Interviewing key executives of manufacturers and store chains, as well as consumer researchers and consumers themselves, you become aware of many concerns regarding packaging issues that are on their minds. We may not agree with all their views or their suggested solutions, but since they surfaced repeatedly in our interviews, they deserve discussion.

Do you buy large bags of chips? If you like them fresh and crackling, think again. You have cereal and chip packages, one retail executive pointed out to us, that are so big that when the expiry dates printed on them indicate a 60-day shelf life, they really have only a 30-day shelf life once they are opened. There is nothing on the package that warns the consumer that once they open the bag, the clock starts ticking and there are only 30 days left for the products to be really fresh. The same might be said about packaging for many other perishable foods.

So, among the issues that we must address more seriously in the future is the need to analyze

When super-large bags are opened, product freshness diminishes more quickly than the expiration dates imply

the excesses of current package sizes and find ways of either modifying them or being more forthcoming about the true value of such humongous packages. It might cramp our style a little, but we believe that shoppers are entitled to be fully and honestly informed about the nature of the products they are buying.

Health and nutritional issues

What do shoppers expect from packaging regarding health and nutrition?

Shoppers are getting more inquisitive about the nature of the true product benefits, and expect the packaging to reflect that factor. They expect the packaging to reflect what they're buying, whether their priority is cost, efficiency, quality or concerns about health. "There's a lot more at stake in the design of a package than just the appearance", explains Eric Greenberg, an attorney specializing in packaging laws.

Health and nutritional issues of food are among the issues that will become more important to consumers. Saturated and unsaturated fat, sugar and salt, listed on packaging, are going play an ever increasing role in helping the customer to make the decision about what product to buy.

Consumers are trying to balance their diets and they want to do it themselves. They need to comprehend the Nutrition Facts and ingredients panels in language that they can understand. They may understand the percentages of cholesterol, sodium, carbohydrates and protein. But little else on the back or side panels of packages really clarifies anything for most consumers.

Frankly, your average consumer doesn't have a clue of what that long list of chemical ingredients really means. So, if consumers don't understand what these chemicals are all about, are the ingredients lists on food packages really helpful?

It's an issue ripe for review. Sticking your head in the sand and ignoring or

Nutritional copy is essential to many shoppers, but do they really understand the implication of all the listed ingredients?

minimizing this issue is not enough. There is a need for working more closely with nutritionists and, indeed, with the food administrations in various countries. Packaging needs to be more candid about issues that relate to product preservation and consumer weight and health issues and in terms that are easily comprehensible to the average consumer.

And then there is the 50+ population

With the rapidly growing population of baby boomers and the elderly, structural package design and graphics highlight the physiological difficulties of this large consumer segment of the world population. Their problems fall into two basic segments:

1. Ergonomic difficulties
2. Visual difficulties.

Pharmaceutical packaging, more often than not, comes into the limelight because the elderly consider many drug packages difficult to open and difficult to comprehend. Products that are difficult to dispense and tiny copy that is too hard to read and too complex to fully grasp its meaning really make no sense to us.

In the book *Brand Medicine*, edited by Tom Blackett and Rebecca Robbins, in the chapter titled Packaging for the Elderly, we discuss how nothing ignites the scorn of the elderly more than their difficulties with tamper-evident and child-proof packaging. Serious conflicts, we point out, crop up between closures that are meant to protect children from harm and closures that are geared to the needs of the elderly.

Child-proof closures naturally try to prevent children, curious as to what is in a package and innocent of the danger of drugs, from package access while the intended product users want easy access. For example, many bottle closures call for pressing down hard or lining up elusive and often hard-to-see arrows, making many packages simultaneously child-proof and elderly-resistant.

Another issue that we need to address is the ability – or inability – of the elderly consumer to read and comprehend critically important information such as usage and dosage declarations, identification of active substances, warnings and counterindications. Even though most elderly people wear corrective glasses, they still find it difficult the read fine print on many packages of pharmaceutical products,

What's good for child-proofing is problematic for many elderly consumers, such as matching hard-to-see arrows on caps and bottles of pharmaceuticals

especially those panels that contain important information on dosage, usage and possible side effects.

No question. Legibility of packaging copy, especially on medical packaging, has always been an important issue, and continues to need greater attention.

According to Mona Doyle of The Consumer Research Network:

Print size and print contrast on pharmaceutical packaging is a major problem. Everything that makes the product easier to use is important to these consumers. Opening and closing are not the only issues. It's easier to recognize, easier to read, easier to carry, easier to dispense without spilling what is important to the consumer.

"The biggest challenge in branding and packaging of pharmaceuticals," Ms. Doyle warns, "is going to happen in larger measure when the first lawsuit is filed by people who feels that some packages that were too difficult to comprehend led to some type of harm to them. This could become an access issue the way the Americans With Disabilities Act is an access issue."

Recognizing elderly consumers as an opportunity

While dealing with the infirmities of the elderly usually elicits yawns by consumer products manufacturers and marketers, one supermarket company in Europe sees it differently and has begun to move in the opposite direction by looking at the elderly consumer as an opportunity, rather than a problem.

In the outskirts of Salzburg, Austria, and in Vienna, ADEG supermarkets cater specifically to the 50+ population. Not waiting for packaging manufacturers to make shopping easier for elderly consumers, ADEG, a subsidiary of the German food company Edeka, features stores that have wider isles, non-skid floors, shopping carts that are easier to handle, benches to sit down for a brief rest and even blood pressure measuring devices. Merchandise displays relieve lengthy searches for products, and price tags have larger letters and numerals. Store lights have less glare, and magnifying glasses, hanging on chains

In consideration of the infirmities of the elderly, ADEG Supermarkets in Austria provide conveniences such as blood-pressure measuring instruments and magnifying glasses

Photos by Haslinger, Keck

throughout the store, can help those with special vision problems. To make it easier for elderly shoppers to unload their purchases into their cars, the parking spaces at the stores are extra wide.

What makes all this even more intriguing is that ADEG has found that the conveniences in these stores have attracted not only the elderly. Half the customers at ADEK stores are *under* the age of 50, being attracted by the amenities of the store in the same way as those over 50.

Now, if that's not intelligent merchandising, what is?

What's good for the goose is good for the gander

The ADEG initiative shows that you don't have to be over 50 to appreciate convenience. Stop identifying the 50+ consumer as a limited market, associated with undesirable and often uneconomical expense. This is equally relevant to packages, whether these contain pharmaceuticals, food, hardware, clothing, or any other type of consumer product. Think of easy opening features and good legibility as an opportunity to gain the goodwill of consumers of *any* age.

Accessing jars of pickles, preserves or pasta sauces with covers that require the strength of a sumo wrestler to pry open and clamshell packages that defy getting to the product without the need for cutting tools are culprits of the same nature. Yet, whether it is food, household products, hardware, electronics and a myriad of other product categories, manufacturers seem be more concerned with packaging line speed than consumer needs. It's economics over ergonomics.

A few improvements

In fairness, there have been a few improvements, but they are few and far between.

Orange juice containers, once sharply criticized for the opening difficulties of the seals on their pour spouts, finally responded with an integral plastic pull seal.

Some blister packs have added perforations to the paperboard backings that facilitate easier access to the products, although the presence of these perforations is often difficult to detect.

A few antifreeze containers, once disparaged for being difficult to balance and apt to spill their liquid on their users' clothes, have changed bottle construction to make the bottles easier to hold and pour without spilling. Applause!

We can't stress it often enough! It is paramount that those involved in package development – manufacturers, marketers, packaging suppliers and designers – pay greater attention to the ergonomic needs of consumers of any age. Easy opening features on packages of any type will gain the goodwill and brand loyalty of consumers of *all* ages – old and young.

It's really a no-brainer.

And what about the environment?

Environmental concerns with packaging have played a major role in the debates between the sharply divergent proponents on environmental issues. We need to pursue a more positive role with regard to environmental concerns by designing visionary packages in ways that will minimize their contribution to garbage.

In contrast to the U.S., plastic bags in Europe and Canada for liquids such as detergents and milk replace environmentally unfriendly bottles

You can learn by looking around the world. In Canada, milk comes in plastic bags. In Europe, some detergents are available in plastic bags with screw caps. Why, in the U.S., are we still filling landfills with gallon milk jugs and large plastic detergent containers?

Even the amounts of ink that are used on some packages could be questioned, their chemicals adding to environmental pollution. Do packages covered with large amount of ink really contribute to increased sales? Will a package using six, seven and more colors sell more product than a carefully designed package with four or five colors? We doubt it.

Among the worst offenders are packages for computer accessories. Some are so overpackaged as to border on the ridiculous. Do products that are packaged in layer upon layer of material really attract more accessories' customers? Manufacturers of these products will tell you that this is to discourage pilferage and that such packages wow the digital aficionados, so why worry about garbage dumps?

Extraordinarily over-packaged computer accessories ignore environmental issues, quoting display benefits and pilferage prevention as the reason

Compare all this to the elegantly designed, two-color, cube-shaped package for Apple computers' iSight. It's a pleasure to see, open and use. It shows that computer products don't have to be over-packaged and color saturated to attract users. We need to ask ourselves: Where is the point of diminishing return?

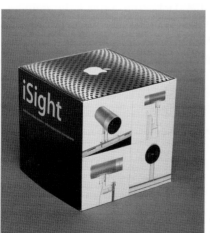

Packaging of Apple products excel in straightforward design simplicity

The third dimension

In discussing issues that will affect future package developments, the potential of structural or three-dimensional package improvements warrants special attention. When we interviewed shoppers as part of our package development assignments, we found universal agreement on one subject. Consumers want portability, resealability and improvements in grip and opening features.

From the package manufacturing side, this suggests numerous opportunities for exploring innovative packaging structures that will translate into packages that will attract consumers and promises increased product sales. It is also an incentive for the design consultancies to take advantage of opportunities for the development of visionary packaging that keep environmental issues in mind.

"Manufacturers, converters and designers of packaging are becoming more and more sensitive to the incentive of creating packages addressing environmental issues, along with consumer demands," states Elliot Young, chairman of Perception Research Services, a consumer research group. "We are seeing much interest in structural package development by marketers and package manufacturers who want to be ahead of the curve. For example, where there are products that currently use plastics or paperboard boxes, companies are now wondering if these packaging structures are the most appropriate. And they are looking at what features could be built into these structures to achieve environmental, as well as consumer benefits."

"Because of the constant change of top management and the drive to increase sales, come hell or high water, there's more risk taking now than there was in the past. Management is more willing to make changes. Companies are realizing that packaging is much more important to their bottom line than they thought in the past."

The time may have come when we need to review our priorities for future packaging. Whether these priorities will relate to better information on packaging, health and nutritional copy considerations, environmental concerns, electronic inventions or structural innovations, or whether there will be repercussions to new methods of retailing, we predict an exciting voyage into the future of commerce in which packaging will play a broader role than ever.

Chapter 6
A broader view

Trying to predict the future has always been a risky business. But if we want to be in control of marketing brands and products, taking an intelligent stab at future packaging is well in order.

Recently, we talked to Lars Wallentin, assistant vice-president, marketing communications and strategic design at Nestlé, headquartered in Switzerland. He travels all over the world for his company and has some interesting thoughts about future packaging:

> Most markets understand the enormous power of the package and of constantly updating their package designs. At Nestlé, we are constantly updating and doing something with our brands in order to make them more relevant, more dynamic. We do it through different graphic design and also structural design. In the world we live in today, if you don't constantly improve on what you have, you will be left behind.
>
> In the future, packages need to be more functional. Packages will be easier to handle – open, close, and yet be more tamperproof. There are new plastic materials which are much more agreeable to touch and structures that are more pleasant to hold. There will be more shaped cans. It's only a question of time and technology when this will be a more common technique. There already are more and more shaped cans in Japan. This is a clear trend.
>
> There will be more bundling of smaller packs. It's more efficient to have a high-speed line with small units that are bundled into bigger units, and also more efficient for smaller families.
>
> There will be more and more material combinations. This gives us the opportunity to make more efficient packaging, combining plastic, paper, metalized foil, or whatever. You could, for example, combine aluminum for lightness, cardboard for stability and plastic for sealing.
>
> Also, people want to have more and more transparent packs. They want to see what they buy. Even Barilla's traditional pasta packages now have windows. We will see more and more of that.

Opportunities galore

"Shipping cartons," Wallentin continues, "are still rarely used as an advertising surface. U.S. packaging is still a brown box market. Look, shipping containers are the cheapest advertising surface there is. Few companies seem to understand that. There could be a big market out there for using shipping containers as a means of brand and product promotion, especially in the hypermarkets, rather than just for shipping and rack displays."

"The cost of printing is very cheap today, thanks to digital printing. There will be more tailor-made packages, even for chains like Wal-Mart or Carrefour. This is still in its infancy, but the minute we go more into digital printing, we can have smaller runs, more tailor-made packaging and packaging made to order."

"There is also a need for more humor in packaging. For example, one of Nestlé's products in England is a chocolate bar, called YORKIE. It's a very thick chocolate slab and it has been positioned as the truck drivers' chocolate. But instead of saying it is for men, it says 'It's *not* for girls', which is a less serious, yet much clearer positioning. And you know what? The ladies are curious enough to try it also.

These bars have become such a big hit in England that Nestlé followed up with another bar, called BLOKIE. The wrap says "It's *definitely* not for girls" and, what's most surprising, the BLOKIE wrap looks almost identical to the wrap for YORKIE, except for the logo. Why? What happened to the sanctity of brand identity?

"The *logo* is not sacred", Wallentin explains, "what is sacred is the *key visual identity* of the wrap. As long as that is respected, creativity can go wild." The bars are a conversation piece in the U.K., and further such unconventional experiments are planned. And when you

Ignoring conventional branding standards, Nestlé dares to project a fun image by changing product names and promotion on some packages while maintaining their trade dress

dare to do things that are out of the ordinary, you get free publicity on TV and radio.

Will future packaging be able to meet all these contradictory targets?

If you put that question to one hundred people, we will probably get one hundred fifty answers.

- Where is packaging really going in the next few years?
- Will global marketing create a need for packaging that is appropriate in regions where customs and lifestyles differ markedly from each other?
- Above all, will the growing audience of Internet shoppers change shopping habits and thereby change packaging visually and structurally?

In our book *Branding@the digital age,* Robert Herbold, executive vice-president and CEO of Microsoft Corporation, comments:

> One of the problems with human beings is that they get good at something and they want to keep doing that same thing over and over again because they are good at it and they don't want to recognize that the world has changed around them. Let's face it, people are good at patting themselves on the back with respect to their skills and they don't want to change these very much.

This may be true to a point, but many marketers are moving in just the opposite direction. Their methods of merchandizing are in a constant state of change, trying to balance the need for making a profit with the need for attracting consumers. These objectives do not always match.

Good intentions – but is it reality?

Wal-Mart, and other mass-market behemoths, often utilizes shipping containers as a means of display in its stores. It claims that one of the reasons for this use of shipping containers is to take the labor out of handling individual packages and instead shift the labor into front end customer service. As well intentioned as this may be, it seems more like wishful thinking than reality. One professional practitioner, experienced in developing store environments, confirmed our concerns during an interview:

Digitally identified product display may not enhance product appearance but supports fast product turnover and accelerates restocking

Packaging really has to tell the product story because there truly is little service anymore at the hypermarket level. Even though these stores say 'Hey, we've got great customer service', they just don't.

People working there have no real understanding, or training, or background in the product mix. So the packaging truly has to not only teach the consumer about what's in the container, it has to teach the sales clerk as well. More and more, the package will have to speak for the brand and speak for the product, simply because there's nobody else who can do that.

There is a plethora of new products, and if you're talking about some hi-tech products, such as photographic products or digital camcorder editing equipment, there's no way that the sales clerk at Best Buy will understand that sufficiently. With the high personnel turnover rate in some of those stores, they just can't train them fast enough. That's why the packaging is really key.

In a study of hyper-outlet chains, researchers watched to find out where the customers looked to gain the information about the product that they were about to buy. Even though manufacturers and marketers may spend millions of dollars on point-of-sale material, what do people do? They pick up the package and read what it says.

Where the rubber meets the road

Advertising and sales promotion are wonderful and they drive traffic. But it's only when people are in the store and pick up the product that the rubber meets the road.

It's the old story: You can lead a horse to the water, but you can't make it drink. You can get the people into the store through advertising and entice them to look for the product. But it's when people get to the product and pick it up that the package becomes the final go/no go decision maker.

It's the last step in the retail psychological pathway to making the purchase.

You have to make sure that that experience is a positive one. Everything is important. It's the store environment. It's friendly and knowledgeable service. It's the ability of the package to communicate a positive message about the product.

The other side of the coin

Not everyone is necessarily enamored with the blandness of look-alike stores that are driven solely by financial criteria at the expense of any interest in the ambience of the shopping environment. In effect, the steel racking approach to store environments ignores any human emotion. In a highly critical article in the *New York Times*, Marshall Blonsky, a professor of semiotics at the New School in New York City, refers to this style of merchandising as "the indifferent equivalence of everything with everything else, for an audience that has no concern for difference and no concernment for quality".

Fortunately, not *every* hypermarket chain follows this track.

Target Stores, for one, try to distance themselves from the emotional black hole of the marketing approach of some competitive hypermarkets. Target features product lines designed by well-known designer and architect Michael Graves. This designer's modern style is applied to a number of Target product lines, from stainless steel cooking sets to toasters and wall clocks, and is displayed in attractive packaging that complements these products.

As in many business ventures, what will count in the end is the survival of the fittest. Will consumers be content with shopping from brown shipping cases in the future or will our shopping environment be brightened by the colorful array of packages to which, until the past few years, shoppers have been accustomed?

Packaging designed by Michael
Graves mirrors the sophistication of
his product designs

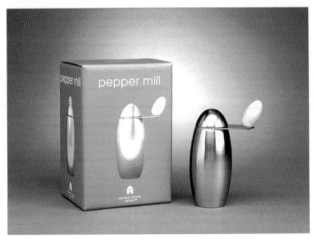

Perhaps we should take inspiration from those retailers who still hold the belief that customers are looking first and foremost for quality, value and then cost.

Rational shopping

When we discuss the retail business with executives of various corporations, we come away with the impression that all that matters to most of them is the mechanics of moving products in and out of the stores with the least effort, the least investment and as little attention to consumers' needs as they can get away with.

All this goes under the heading of success and encompasses everything from the business curriculae at universities and colleges to the desks of corporate business enterprises.

The success in the economics of commerce by merchandising goliaths such as Wal-Mart will, no doubt, serve as the shining example for many marketers and students of marketing.

But the indifference of most hypermarkets to the rational side of human life is alarming to us.

We may be idealists, but we believe that a mindset, in which money is the sole priority in what we do, will not necessarily lead to a satisfying life for most people. There has to be some rational involvement in the things we do or we will be like zombies walking through life without feelings.

Visionary packaging – the challenge

In the next few years, designers will need to face the challenge of calling attention to the need for aspirations that transcend the current preoccupation, bordering on hysteria, with percentages of profit margins.

This is not to disregard the need of any commercial enterprise to be successful in making theirs a profitable undertaking. We all want to live well. And being successful, whatever our profession or business, is a laudable ambition in modern life.

But it behoves corporate executives in the retail business to consider what Marc Gobé, a designer of brands and retail environments, calls in his book *Emotional Branding* "the art of accessing, with intelligence and sensitivity, the true power behind human emotions".

"In this hypercompetitive marketplace," he continues, "where goods and services are no longer enough to attract a new market or even to maintain existing markets or clients, it is the *emotional* aspect of products and their distribution systems that will be the key difference between the consumers' ultimate choice and the price that they will pay…how the brand engages consumers on the level of the senses and emotions; how a brand comes to life for people and forges a deeper, lasting connection."

"This means that understanding people's emotional needs and desires is really, now more than ever, the key to success. Corporations must take definite steps towards building stronger *connections* and *relationships*, which recognize their customers as partners. Industry today needs to bring people the products they desire…through venues that are both inspiring and immediately responsive to their needs."

When all is said and done, shopping without emotion, plucking products from brown shipping cases stacked on industrial racks, with graphics that are reduced to jet-sprayed brand and product information, may move a lot of products and please your investors. But if this becomes the normal approach to retailing and packaging, it will make our shopping experience seems like a nightmare that no one, not even the hypermarket executives, should wish on anyone.

Chapter 7
Emotional purchasing

It is often your emotions that make purchasing decisions for you. Many of the products that you buy and use every day were not bought only for their functional benefit but because something grabbed you from inside and said you would like it. It is often the package that creates the emotion, and the package that reminds you of something special you want to remember.

Marc Rosen, the well-known designer of fragrance brands and packaging, told us about a chat he had with a flight attendant on an airplane. The flight attendant asked him what he did and he replied that he designed perfume bottles and cosmetic packaging.

"Oh", she said, "you know my mother collected perfume bottles all her life and when she died recently she left me all the bottles in her will. I'm so thrilled to have them because I think of my mother when I see those bottles, and now I'm adding bottles to the collection which I plan to leave to my daughter."

Were the bottle designers visionary, knowing that the bottles would be passed down the generations? Probably not, but they did know that the bottles could stir emotions.

Today you can buy a wide variety of decorative and perfume bottles at auctions and antique shops and use them as collectibles. People want these bottles because of the emotional response they evoke. The thing you notice about these bottles is how their shape, or their label, like a Rorschach test, reminds you of something positive or negative.

The silent salesman and emotional purchasing

When you go to a store, especially one of today's large "box stores", and you see this bazaar of products from which to choose, the package must speak to you so that you will reach out for it. It must stir your emotions. It is the silent salesman.

The package subliminally says something to you about its status and implies a level of taste and quality, working hard on the shelf to get your interest.

Ask three women who you know what perfume they wear and why they like the package. You'll find their answers reflect their personalities.

"Burberry Brit" gives a status message very different from "L'Or de Torrente", and you will never find the same person buying both of them. When a woman takes out a lipstick in public, if it's a Chanel lipstick, it says something about her status, level of taste or quality of life. Cosmetics are emotional, and cosmetic packaging is emotional.

Modern British style, with a playful and spirited attitude, is communicated through the bottle shape, the graphics and the Burberry Brit name

The dream of elegance, sophistication and femininity is suggested through the decorated, engraved gold leaves of L'Or de Torrente's oval bottle shape, evoking the sensual curves of a woman

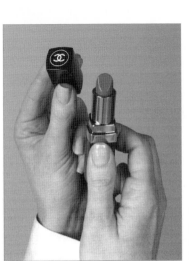

The bold Chanel logo on the simplest of black shapes suggests confidence and status to the user of the lipstick brand

If someone buys and likes a product or the fragrance, they will buy it again. They may not buy only because of the package, but since they like the product, they will remember the look of package.

What makes them remember this package and what stirs their emotion? People always say, "I like the red one". Well, it's a lot more than the color that attracts. If someone has seen an ad for a fragrance bottle that looks like a fan, they might not remember the name of the fragrance or the color, but they will say, " Where's that bottle that looks like a fan?" That unique shape has stirred an emotion and caused an indelible impression.

It's hard to forget this bottle's fan shape, which also reflects the fan-shaped symbol used on Karl Lagerfeld products

Photo courtesy of Marc Rosen

The bottle becomes a defining object much like a car or a wristwatch. The Hummer car is not for shrinking violets, and the lawyer who wants sympathy for his client will not wear a Rolex wrist-watch. The bottle also is part of a wardrobe and it defines them.

You probably know some middle-aged people who have been wearing the same fragrance since they were teenagers, and they are reluctant to give it up. To them, it's their defining object. In fact, these are the people who write letters to the company when it changes the bottle or the label. As a good marketer, you want people to become not only attached to your product but also to your package.

High-quality emotion

A visionary package will imply the quality of the product inside. With cosmetics, a heavy glass or very heavy plastic implies quality, giving products like nail enamels and lipsticks a floating appearance. This is a trend and it will imply that the product is pure, of high quality and looks expensive. Why is this important? Today, people are careful in spending their money but still want a luxurious feel to their cosmetic purchase.

For years designers liked to use heavy glass because of its luxurious impression, but it was expensive. Since glass companies today are more proficient at controlling the costs, the cosmetic companies have the vision to spend the money on heavy glass because they realize it stirs the consumer's emotion.

Changing emotions

Marketers have convinced many women to wear a wide range of fragrances for evening, daytime, winter, summer or whenever, depending on their mood. And the cosmetic industry has tried to market fragrances based on a positioning of being sexy, aspirational, spiritual or whatever the mood is at the moment for which you're trying to get someone to wear the fragrance. A fragrance, remember, is the last accessory, almost like a piece of jewelry.

The ad creates a fantasy of water, florals, beauty and wind, all circulating around the bottle shape

Experience
ESTÉE LAUDER
beyondparadise
Open here. ▶

ESTÉE LAUDER
beyond p a r a d í s e
an intoxication of the senses

The most successful cosmetics ads feature the package. In branding the product, the package becomes the image for the product. What is the first brand you think of when you think of lipstick? Revlon? They certainly want to be first on your mind, and must realize it is important to have the best-looking lipstick case in the industry.

The fun part

Designers of packaging will always want their clients to have the best-looking package in their industry. And in industries where emotions play a large role in selling and buying, the marketers of products always seem to enjoy the packaging part of the marketing process the most.

Again let's take cosmetics. "The fragrance itself is ephemeral, but the package can be held in your hand and it can be caressed", observes designer Marc Rosen. You can like it or hate it, and marketers enjoy giving their opinions about how much they like or hate the packaging when it's in its design stage.

"Beauty without expression tires", wrote Ralph Waldo Emerson. To create emotion in a package, today's designers can't just think of designing that pretty shape. All the pieces of the puzzle have to come together to eventually stir emotions, and this could be the product name, the fragrance concept, the cosmetic product and even the public relations associated with it. A package with vision comes from a designer with vision, and you need to understand all the issues associated with marketing the product. Those products that succeed will always have consistency between the name, product, package, advertising and public relations.

The male emotion

We speak a lot about women reacting emotionally to packages, but what about men? We know that the man thinks of his car or wristwatch as being his defining objects. We also know that today, in grooming and in toiletries, the package has become a defining object for men.

Think of men's toiletries and some of the images that are associated with them. If you go into a department store, or even a drugstore, you'll see men's toiletries and cosmetic products with shapes resembling automobiles and even the parts for automobiles – grilles, nuts and bolts, spark plugs – all these shapes have been used for toiletries with the vision of evoking the male emotion. The Gillette lines of

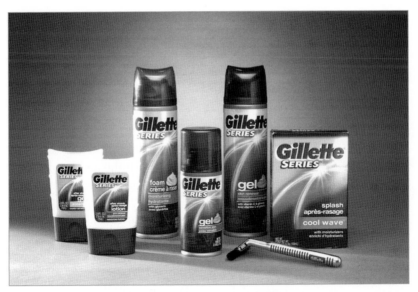

Gillette is experienced with masculine touch points. The combination of ribs on bottles and razors, and silver color suggests industrial metal and automobiles

men's toiletries and razors are examples of mainstream products with the emotions and shapes associated with industrial metal and automobiles. The lines' consistency are also part of their success.

And how important is the name? Think to yourself of the various products you could apply the Porsche name to and the emotions that would result from it.

What does a car have in common with a package? Several years ago we were trying to develop a series of designs for an Old Spice line of men's products using the sports car theme. We photographed sports cars from all angles, and used outlines of grilles, air intakes, hood ornaments, lamp mounts and even hubcaps as inspiration for the package shapes. At the end, the auto theme was not selected for final development, but working with those elements stirred an emotion among the designers, and we'll never know how successful it would have been.

Don't think emotional packaging is confined to cosmetics. Product designers package the inner works of products in shapes to evoke emotions, especially in stores like The Sharper Image. There we see electronic and technology products, such as cellphones, along with non-electronic products like clocks and suitcases, still evoking a technology motif and the emotions associated with technology.

Facconnable men's fragrance, a popular store brand from the Nordstrom chain, uses technology cues for its bottle shape to appeal to the electronic-toy-loving male population

Packages for products such as Perry Ellis Portfolio and Aramis Life take on the feeling of these electronic products and the emotions associated with them. Although the automobile was always the benchmark of the day for design, today's styles come from visions of TV's, laptops and cellphones, even more so than automobiles.

There are many types of stores in which you will see packaging that evokes strong emotions, starting with the everyday grocery store. Seasonings such as Tabasco Sauce come from companies that understand that even though these are commodity products, they can sell more through the emotional attachment of the bottle shape used at the table.

And the liquor store is a great example of emotional packaging. Going back to one's defining object, they say that what you drink is what you are. The liquor bottle often reflects the brand itself.

Without the recognizable bottle and label of Tabasco Sauce sitting on the table, it would be difficult to enjoy your meal if you like spicy flavoring

A quick test was performed where a group of about 50 marketing people were shown 10 silhouettes of liquor bottles. These were black silhouettes on a white background with no labels of Crown Royal, Tanqueray Gin, Courvoiser, Absolut Vodka, Jack Daniels Bourbon, Kahloua Coffee Liquor, and others. Can you identify the brands? Would you believe that practically 100% of the group was able to identify all ten products. This not only shows how young adults enjoy the taste and romance of spirits, but also how the impression of bottle shapes have been embedded into their minds and their emotions.

It's important for you to realize that both sexes are affected by emotional packaging. We forget that only a few years ago, men would borrow their wife's moisturizers and toiletries, but now are finally secure enough to buy their own.

Can you recognize all four liquor brands by their bottle shapes? (Crown Royal, Tanqueray, Courvoiser, Absolut)

Visionary packages stir the emotions of people all over the world. They may have slightly different connotations in different areas, but they are becoming more closely associated with one another. For example, the impression that French fragrances must be better than the others does not exist any more. Brands like Calvin Klein and Ralph Lauren are popular worldwide and have not only been visionary in their advertising, promotion and packaging, but have shown the universality of the emotional link to the global product.

Chapter 8
The lifestyle
influence

With its sour taste, it's amazing that people can enjoy eating yogurt. Rarely will they order it in a restaurant, but they will always buy those small portable, serving-sized cups that are divided into portions and then thrown away.

The yogurt people were visionaries. Several years ago they saw that their consumer of the future was living a mobile lifestyle and their product concept was centered on this mobility. They took advantage of today's consumer who eats meals everywhere – at home, in the office, on the road, and in between meals. So by understanding consumers' lifestyles, they built a worldwide industry around a serving-sized cup.

This visionary yogurt package is so influential, that those who find yogurt only marginally acceptable as a food, accept its taste because of the package convenience. This is a good example of a package that takes advantage of lifestyles – mobility, diet consciousness, simplicity and convenience.

Even the live or perishable products are changing in their nature. In the old days you bought your cantaloupe, took it to the checkout counter, paid for it and it was put in a bag. Tomorrow's cantaloupe which will be precut, carved and mixed with other types of melon, fruit, ginger sauce or whatever will require a new type of package. These new products and their packaging relate to lifestyle changes.

So portability isn't for just getting the product out of the store, but also for ease of handling and even eating in the car.

Have you bought any water lately? Would you believe that bottled water is the fastest growing product in the beverage category? According to C. Manley Mopus, by the year 2010, bottled water will be the leading beverage, surpassing soft drinks and juices.

Who would have ever believed that we would be importing bottled water from all over the world, and selling it at higher prices in some cases than many fruit juices? Just today we must have seen at least ten people carrying bottled water with them while going about their everyday business.

You need to keep pace with the new mobile, grab-as-you-go, lifestyle. The food industry understood this lifestyle when it brought out products like cereal bars with milk, hand-held soups in disposable, microwavable cups, and tortilla chips that are packaged with dips where the lids turn into bowls. All these open new challenges for new innovations in packaging. And the product itself may not be new, but the package will make the difference between eating at the table, or eating on the run.

Do you eat with chopsticks? People are traveling more and ethnic foods provide you with many challenges. You have the opportunity to design packages that have ethnic-inspiring designs. And the stores themselves can treat their shelves and aisles like destinations within foreign countries.

In fact, marketers and designers can take advantage of the information they gather from imported products from all over the world, understanding color and technique and other attributes that appeal to ethnic consumers within these parts of the world.

How do we design for today's lifestyles?

Let's say that you are the marketer of a new beverage. But there are hundreds of beverages already out there. So where will this beverage fit? If you understand the consumer and can pinpoint how this beverage will fit within their lifestyle, you have a better chance of success.

You need a vision and you need to delve into ongoing trends. What are the choices that surround this new beverage? Do we drink it cold? Do we drink it hot? Is it about drinking during relaxation, or drinking for more energy? If it is about energy, what is this world of energy? And what is the consumer's lifestyle?

According to Pam DeCesare, director, packaging and brand design for Kraft Foods, from the beginning you need to focus on the consumer – what they see, what they think, what they want, don't want, and what they feel. You need to take this information and integrate it into a visual expression so that the designer and the marketer can see the things the consumer is seeing and can understand which of those things communicate and which do not. The brand opportunity for the new beverage and its vision for the future will be in defining what communicates.

Research should be brought into the branding procedure. Marketers can go to the consumer and discuss different approaches and different experiences in order to understand how best to communicate their brand.

For example, let's take coffee as this new beverage. Coffee gives us different feelings. One may be a sensual kind of coffee experience. Another might relate it to jet fuel – that spurt of energy and activity. Another might be having a very relaxing moment with coffee. Consumer lifestyle information coming out of these questions enables us to position the new product. There may be certain color cues, or type styles that get us to that positioning.

So even before the design study commences, these questions are asked in order to understand the brand. This enables the design phases to be less subjective and work more quickly. Everyone is now on board, and analytically everyone understands what should be done. This pre-design information – and I think of it as lifestyle information – is a look that talks to marketers and designers. It gets them to talk to each other and create a visual language. According to Pam DeCesare, this approach was a big change for Kraft from the method of picking from a series of designs by consensus. This is a focused approach and is the way of the future.

Lightening the load

People like to think of ways to simplify their own lifestyle. Recently there was an article in the *Wall Street Journal* about an army sergeant who had come back from Afghanistan and had written to his superiors about his experiences. These were not battle experiences, but ones involving everyday matters. He spoke about the food packages and the ready meals that army personnel take with them in their huge rucksacks, and how these packages weigh over two pounds each. He said that when the soldiers get to their destination, they throw away everything except their granola bar and some basic protein elements. Everything else is thrown out so that they can move around more easily.

The email written by this sergeant went around the army ranks like an explosion. All of a sudden this young sergeant was being interviewed by people in charge of army supplies for his vision of what their packages should be like. The army took these suggestions very seriously because they know how important it is to lighten the soldiers' load. They saw the opportunity to simplify and lighten the packages that the soldiers carry. And you know that lighter and more portable packages will also connect with civilian lifestyles in the future.

Keeping it simple

The idea of keeping things simple is not new. Things around us started to appear more visually simple many years ago. At the beginning of the 20th century when decoration in art, architecture and products became difficult to produce, the Bauhaus came along in Germany, and made those elements of design, especially in architecture, become simple to manufacture. Those simple architectural lines became the look of the new structures that started with the Bauhaus. Art forms became more simple and less decorative, and products themselves, in the early part of the 20th century, started shedding their strong decorative elements. Soon typography, advertising and packaging took on a more simple look and the flourishes and decorative elements that had been used between the 19th and early 20th century changed to less complex forms.

What will be the next revolution in simplicity? The opportunity seems to be in making communication more simple for our busy lifestyles. Banks are now advertising for customers by saying that their forms are easier to fill out. You can apply for a mortgage more quickly and simply. Insurance companies are trying to do the same thing.

We now have a great opportunity for the packaged goods industry and the retail industry to make lives simple. Who would have imagined a pre-made peanut butter and jelly sandwich? Smucker's, the jam and jelly company, has this product and it's become a tremendous hit. Was Smucker's visionary? Actually, Smucker's was reluctant to put it out, thinking that people should certainly have the time to slap peanut butter and jelly in between slices of bread. But once they had found out, by asking questions, that many busy people just don't have the bread on hand to make the sandwiches, and can't always rely on someone being home to make the sandwiches when they are needed, the frozen peanut butter and jelly sandwich was born. An example of the ultimate in portability.

Keep it simple. By emphasizing the shape of the product with the familiar Smucker's plaid background, you might be convinced of its convenience. The white plate should have been left out to further convey portability

Simplicity in shopping

Simplicity is harder to achieve than complexity. We get messages from advertising, stores, the Internet and our neighbors – how do we make decisions? The big opportunity will be to enable consumers to make purchasing decisions through packaging which is simple, easy to find and easy to understand. Packaging where less is more.

With today's lifestyles and people being in a big rush, shoppers in self-service stores often get frustrated. The helpful service is not there anymore. But the successful stores know how to simplify things to relieve frustrations.

A good example of this is Best Buy, who understands consumers' frustration in spending time looking for products that are difficult to find. Its signage is simple, its color schemes are easier to understand, and part of its mission is to make customers' lives simpler by making it easier to shop. This understanding of the consumer puts Best Buy among the hypermarket exceptions. The opportunity comes with also balancing service personnel with self-service so that you can visually locate what you need without spending extra time.

It's easy to shop and find your products at Best Buy, and customers can check out packaged electronics products like they do with groceries at supermarkets

Many people lack boundaries in their lives, and are often overwhelmed and cannot make choices. But simplicity will give people the freedom to control what they do with their lives. That story earlier of the young army sergeant in Afghanistan foregoing his meal ration, indicates how someone with very little power in an organization can take it into his own hands to simplify things and be rewarded. This is a good example for today's organization, which realizes that the general public can tell marketers how to make things simple for them.

Where do we put it?

Changing lifestyles will bring new products onto the market, but some of these new lifestyle products just won't fit into already established store categories. This is probably one of the biggest challenges for marketers in the future.

Keep in mind that the package can work differently within different channels of distribution. For example, a product package works against different parameters in a grocery store than it does in a nutritional center or health club. So new products can be developed that don't fit into already established store categories. There is a big opportunity to design the package and be the first to set the standard in a category before the competition rushes in.

A recent category entry is the Balance Bar. The package makes it look like a candy, but the marketers tell you it's healthy. Is this product a chip type of snack, a cookie type of snack, or a candy type of snack? What is it? With new products like these, we blur the categories and blur the channels in which the products and their packages will live. This is another big packaging challenge for designers, especially because of the consumers' changing lifestyles and changing what they do.

Lately, the stores have been bombarded with packaged ethnic food varieties – Chinese, Japanese, Mexican, Cajun. Walk into your local supermarket and try to find the ones you want. It's confusing. Some stores actually have an ethnic foods section. But often the products are scattered with the rice and sauces and so forth. The challenge is for designers to billboard the packages for easy spotting.

Changes in the store

Have you noticed when shopping in the supermarket that there is a sexy part of the store and a dull part? The duller items have always been placed in the center part of the store. This includes cereals, cookies, pasta and such, whereas the sexy products in the periphery of the store include the meat, fruits and vegetables. It's the living products versus the preserved products.

It is possible that in the future the largest percentage of these center store product orders will actually come through the Internet.

What's a store to do? One possibility is to change the type of packaging that has been traditionally used in the center store. Change the shapes of boxes, bottles and cans that we have seen for decades. Ready meals, single-serve and other new lifestyle-type food products will become more important, and packaging will have to keep up with those lifestyle usages.

The visionary commodity

The popularity of gourmet cooking has increased. Although more people are inclined to spend less time cooking their everyday meals, those special meals and guest meals have increased when associated with gourmet cooking. But vinegars, sauces and many cooking products were considered commodities.

This commodity attitude applied to packaging as well. Heinz packages were typical. Heinz wine vinegars for cooking originally used a basic bell-shaped bottle with a metal screw cap closure and a bar-coded front panel label, which made the product look cheap and ordinary.

When we realized that these wine vinegars were more often being used for gourmet cooking along with ordinary recipes, it seemed like a good opportunity for serving the gourmet market.

Without an advertising budget and without incurring major expense, our vision was for the brand's packaging to be the accelerant to deliver the upscale message.

First, the bottle needed improvement. Retaining the basic bottle shape, the opening of the bottle was changed at minimal cost, and a gold-colored plastic cap replaced the metal screw cap. This achieved an alignment that permitted the use of a wraparound paper neck label, which, together with the gold cap, gave the impression of a wine bottle.

This was a breakthrough for Heinz. The vision brought several design revisions over a period of years resulting in many improvements, including an enlarged label and unique label shapes resembling premium cigar bands, primary emphasis on Heinz identification, gold borders, rich colors and ribbons, creating an upscale gourmet perception. Product-related illustrations added appetite appeal. Labels placed on the back of the bottle added some clever advertising copy to the mandatory information required on all labels.

The vision became a reality, and the improvements of bottle design and label graphics enabled Heinz to successfully relaunch five wine vinegars as the only national brand in the gourmet vinegar category.

An upsurge in sales of the product line followed this upsurge in gourmet usage.

The original bottle showed all signs of a commodity product. By aiming toward the gourmet market, the new bottle shape and label widely increased the appeal and sales to the full range of consumers

The visionary snack

Our kids are following our mobile lifestyle and they're eating on the run. They too eat their snacks in the car and at play.

Nabisco Brands, the leader in the cookie and snack cracker category, had a vision of developing a totally new brand, which would be based on a cracker character to appeal to a broad range of consumers. Nabisco had the ingredients and the facilities and believed that its Graham crackers could be used for the taste to appeal to this range.

In order to develop a character that encourages hand-to-mouth snacking, we did a major concept study using many types of people, animal and cartoon character shapes. With the vision of marketers and designers, little bear-shaped characters were identified as timeless and appealing to all age groups. So bear-shaped cracker products were designed, using a wide variety of facial expressions and positions.

While the little bear shapes were developed, the package graphics were also designed to convey a strong personality. The package showed these little bears bursting out of the package itself. The product quantity shown on the package enabled the consumer to view these cuddly teddy bears. This encouraged snacking.

With many flavors added to the line through the years, the Teddy Graham shape has become one of the most successful food product

It's a winner, since the bear shapes appeal to all ages and the taste is derived from Graham crackers. The carton graphics emphasize the cute and portable little bears

launches in U.S. history. Through the vision of the marketer and the designer, this ubiquitous little bear has become an icon for food snacking throughout all age groups.

The visionary tea bottle

Try a taste test. Take three glasses and fill each with a different brand of iced tea, but all with the same flavor, and see if you can taste the difference. Odds are you won't be able to taste the difference among the three brands.

Why is iced tea anything special? Whatever the reason, iced tea became a major lifestyle beverage in the late 1980s, and the Snapple brand took a big lead with its fairly heavily advertised flavored bottled tea. Many other brands followed suit. In fact, Snapple became so popular that it sold its company and product line to Quaker Oats for what became an overpriced sum. A few years later, when Quaker Oats realized that Snapple was not as successful as they had hoped, they resold it at a huge financial loss and fired their CEO for buying the company in the first place.

Throughout this time, a visionary company named Arizona Beverages was producing tea. It started in 1992 with a 24 oz. can, and later that year packaged the product in glass bottles.

Why would you buy Arizona's product? The big difference between Arizona and the other iced teas was its package design. In 1994 Arizona introduce an Iced Tea with Ginseng in cobalt-blue bottles. The bottles were well designed and the product instantly became a top-selling flavor for Arizona, selling over 20 million cases that year.

Throughout, Arizona has relied on its package to entice the consumer and make the sale. If you are a bottled tea drinker, you'll remember its green tea products and their memorable label overwraps.

Tea is tea, but Arizona's big differentiator is the appeal of the bottle to the aesthetic sense of the consumer

Partnering with artist Peter Max, Arizona also brought new "limited edition" designer packages to the market in 1998. This was another visionary package approach to heighten its sales.

Arizona has concentrated on its packaging to differentiate strongly from any other brand. It has used innovations in the category such as a grip glass bottle for the Lemon Tea, and established unique flavorings by partnering with Celestial Seasonings and Allied Domecq's Kahlua Coffee. Arizona was the first to introduce a plastic long-neck bottle, which had never been used previously in the industry for hot-fill beverages.

Their unique RxHerbal brand won the prestigious London International Design Award for "Overall winner in package design" in the year 2000.

You could argue that the growth of their company has been secured through their vision of packages that are memorable and, in some cases, even collectible.

The company recently dived into the herbal-enhanced flavored water category. This category already has imported and domestic products, which appeal through their package design, so Arizona has a tough challenge in store.

The visionary wine

Everyone thinks they know something about wine. Most of us know that wine comes from grapes, and that the better the grapes, the better the wine. In some parts of the world, the best grapes are a result of the climate. In other parts of the world, the best grapes are the results of the soil.

In Argentina, in the province of Mendoza, the concept of altitude is the determining factor in obtaining the best grapes. The wine producer, Chandon Argentina, determined that the height above sea level in Mendoza generates an effect similar to the cooling effect of the ocean, with temperatures that vary substantially between day and night. This key factor enables each wine variety to reach its optimum ripeness.

Chandon grows its grapes on terraces located at the foot of the Andes Mountains in Argentina, in a series of elevated zones from 600 m to 1,600 m above sea level. Each zone has a distinct climate, perfect for grapevine cultivation and eventually for the production of good wine.

In marketing this premium wine and working with the designers at Interbrand in Argentina, the vision was for the wine bottle and label to reflect the altitude from which they are grown, and the terraces on which they are cultivated. But how can the packaging of wine indicate that the product comes from different altitudes in the Andes Mountains, and still reflect the premium imagery needed to complement the brand?

First, a brand name, *Terrazas* meaning "the terraces", was generated to reflect the concept.

The brand identification was accomplished through a new label design, with deep brown colors that relate to the mountains. The label becomes an extension of the mountains, the land and the Amazon. The labels show the altitude from which the grapes are grown, and each label design extends the landscape of the Andes Mountains. The brand stands out from the usual white, fresh-looking labels used on wine bottles in the U.S. and Europe.

There are four red wines and two white wines. The backgrounds of each label use earth tone colors, along with illustrations of the mountains.

In order to further enhance the vision, a metal canister is used as an outside wrap for each bottle. These canisters enable the label to be more explicit of the altitude, and emphasize how the mountains surround the viewer. The wine varieties – Cabernet Sauvignon, Malbec and Chardonnay – use different height mountain illustrations to point out their different altitudes to the consumer. The canister also makes a premium gift package.

This new product line has become Chandon's largest brand outside France. And this successful product will eventually expand to the U.S. and European markets.

The vision of emphasizing the altitudes from which the wine is produced has enabled these premium wines to be memorable throughout Argentina and eventually in other parts of the world.

Terrazas wine builds its quality perception by emphasizing on the canisters the altitudes at which the grapes are grown

The visionary bakery

There are some products that are part of a packaging category that has been the same for decades. Take bread. A shopper would hardly consider paying extra for fancy bread packaging and therefore, in most parts of the world, the paper overwrap, or polybag overwrap, is universally used on bread. What consumers believe is that bread is bread.

Hovis Bakery in the U.K., with a reputation for good quality and good taste, had for many years used plain packages. Its standard logo appeared over various solid colors to represent the different types of bread.

But Hovis Bakery had the vision of capturing the imagination of the public.

How can the consumers' perception of bread change from a dull, everyday product to something more and imaginative? Hovis took a radical step.

Instead of showing the usual pictures of bread, grains, ovens, old-fashioned bakeries, or other visuals that you sometimes see on bread wrappers, Hovis used photographs of things that U.K. consumers love to spread on top of the bread. For example, the "Great White" breads have wraps with close-ups of cucumbers. Some "Great White" breads have wraps with close-ups of baked beans. "Brown Bread" uses cheese photographs and "Hearty Wholemeal" bread uses tomato photographs.

The vision was to show the experience, reflecting the consumer's lifestyle and bypassing the actual product statement. Designed by Williams Murray Hamm in London, the packaging has won many awards.

When the new packaging was introduced, Hovis embarked on a teaser campaign involving actors wearing "baked bean" patterned suits, and extensive press and TV coverage. This demonstrates the way packaging design can be used to completely reorient the position of a brand. Hovis went immediately from homely and old-fashioned to progressive, and its sales increased over 20% in the year following the relaunch.

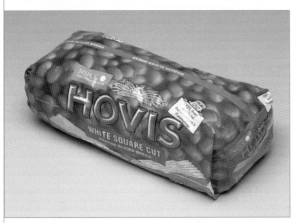

The bread wrapper photographs enable Hovis packages to stand out from anything else in the category. Perhaps this changes the public's attitude about the dullness of bread, and is a response to the biblical slogan "man can not live by bread alone".

You need to stand out in a dull category. If baked beans are associated with white bread in the UK, why not flaunt it?

Lifestyles aren't always global

In some countries there is less space for storing packages and products. This can be bad and good. Bad because of the inconvenience, but good because it has produced higher standards of food packaging and these high-quality standards have led the way to new innovations.

Let's take Europe. The innovation in packaging equipment has not come from the U.S. It has mostly come out of the European market. The Europeans have had more environmental issues to deal with than the Americans, and Europeans have had to work on their sizes of packaging to make things fresher and last longer.

Asia uses other techniques to move products. Many stores in Asia change fixtures, racks and merchandising several times per day in order to encourage their breakfast, lunch and dinner customer. These are issues that are important for the global retailer and producer to understand when developing packaging that will be sold in different parts of the world.

A vision of authenticity

We've talked about the proliferation of ethnic food, but how do you certify its authenticity? What it comes to ethnic food, consumers usually want the taste to be authentic. With lifestyles changing around the globe, consumers are also looking for a convenient way to prepare their meals. How do you combine authenticity with convenience?

A product line that does this is Asian Home Gourmet, the first company to take Asian foods out of the ethnic Chinatown and successfully position the brand in supermarkets worldwide.

Kwan Lui, who is of Burmese origin, really missed the taste of Asian food when studying in the U.S. When she returned home over the holidays, her cousin had gathered together some packets of spice pastes from his army rations for her. This contributed to her idea to develop a business manufacturing spice pastes and introduce the delights of Asian cuisine to consumers all over the world.

A new brand of spice pastes was born. Originally the brand, Asian Home Gourmet, was distributed in over 20 markets globally, and positioned differently in every market. Each market had a different package design, dictated by the distributor at the time. What was the result? Low brand recognition on an international scale.

The brand marketers and the Interbrand designers recognized early on that for this Asian brand to gain awareness and scale in markets outside Asia, the brand had to be in mainstream supermarkets and not just ethnic outlets. Today's consumers expect more exciting cuisines and authentic products on their supermarket shelves. And packaging would be vital to this vision.

The designers defined and developed a consistent global brand proposition for these Asian products that would not alienate non-Asian consumers but, at the same time, was believable to consumers in Asia. The brand proposition revolved around Asian values and authenticity, coupled with convenience and creativity.

The Asian Home Gourmet product line includes spice pastes for rice, meat, curry, soup, noodles, poultry and stir-fry. But how can the packages emphasize the authenticity? The designers developed package graphics and styling to reflect the respective area of origin – Indonesia, Thailand, India, China, Singapore and Vietnam – of each of the products. The Indonesian products show photographs using charcoal burners. The Thai products have their distinctive ceramics. Indian, Chinese and other products are each represented by the patterns and the vessels used on their packages. The authentic-looking Asian packages show patterns and vessels actually used and recognized by Asians in their regions. Every element on the photograph of each package promotes the products' authenticity. The food is authentic and the packages emphasize that attribute.

The designers also developed a global packaging system that highlighted the pan-Asian offering by color coding the different country cuisines. This helped to differentiate the Asian Home Gourmet brand from other ethnic brands. Each product category is retailed at a different section in the supermarket for greater brand exposure. For example, the meat marinades are sold at the meat counter, the rice pastes are sold at the rice section and the salad dressings are sold at the vegetable section.

In order to visually hold the brand together worldwide, the logo, graphic format and type fonts are used in a consistent manner throughout the product lines.

Packages were designed for today's modern consumer, with appetite appeal, ease of use and authenticity as the key elements.

To carry through the theme of the brand on its website, the packages are used as icons to display the recipes and emphasize the product line.

And the packaging is only part of the vision. Asian Home Gourmet reinforces the brand with tasting sessions in supermarket aisles, which is a common occurrence in Southeast Asia. It builds a customer database through magazine subscriptions, recipes and its *Entree to Asia*, a six-part TV series featuring a western chef producing delicious, authentic Asian cuisine. The company publishes a magazine, which focuses on Asian cuisine and lifestyle as part of the program. This program is also broadcast weekly in Canada and the U.S. on public television. And the consumers appreciate the homemade taste of the products.

The result of the visionary program is an Asian brand combining lifestyle and convenience with authenticity. It is made in Asia by Asians, but competes with other mass-market consumer brands worldwide. Emphasizing the Asian authenticity, the company has grown and the products are available in over 6,500 major supermarket and chain stores.

*The background patterns and the vessels used
for the photographs are indigenous to their
regions to communicate the product source
and authenticity*

The convenience factor

It's convenient to shop at the store that's two minutes away from where we live. But people drive for many miles to what they consider more convenient than geography – the convenience of easy shopping.

We will go to stores where it's easier for us to find what we're looking for, and easier to know what the prices are. And the stores are trying to make things more convenient for us. The opportunity lies where the store can create a more instant communication between itself and the shopper.

And today's consumer believes that packaging can make life easier and reduce hassles. If a package is portable and makes a positive difference in their lives, they will buy the package. It's not only the products that make lives easier, but the packages that hold these products make it easier to use and easier to carry.

Look at the home cleaning section of the retail store. The spray bottle itself was a visionary thing years ago, which now is part of our everyday lives. Consumers said that this convenience is important and they are willing to pay for the convenience of spraying rather than pouring and mixing in a bucket. It's amazing how many trigger-spray bottles there are out there.

The convenient pour

You're driving along, the heat gauge in your car registers high, and it's overheating. Your car needs water – no, it actually needs antifreeze. You don't think about that bottle of antifreeze until there's an emergency, especially on the road.

For years, antifreeze was sold in big plastic bottles, which were almost impossible to aim under the car hood and pour into the designated radiator. The bottles had a handle on top and required a contortionist to aim. When pouring, the liquid would "glug" – it would pour, then stop, then pour, then stop – and go on until the engine itself was soaked and the driver was frustrated. Although these plastic bottles were never convenient, they were traditional and every antifreeze company used a similar bottle into the late 1980s.

Our client, a chemical company, decided that was time to make life easy for the driver and simplify the system of pouring. The vision was to make a better package, which would be easier to handle, aim and pour.

With much conceptualizing and design, we arrived at a method for displacing the air that rested above the liquid inside the bottle. This would eliminate the glugging effect.

The next step was to develop a bottle and handle shape that fitted comfortably in the hand and also had the ability to pour easily when the car hood was open.

Instead of putting the handle at the top of the bottle like all other antifreeze bottles, a visionary step was taken. The handle was aligned with the spout on the side of the bottle, which combined air displacement along with an aim mechanism for effectively pouring.

The client immediately realized the convenience this bottle would offer the

There's a solution to every problem.
Aligning the spout with the handle made
pouring simple and reliable

consumer, and promoted it heavily as the "no-glug jug". Every company producing antifreeze wanted to get hold of it.

So this patented, visionary bottle changed the way liquids were poured from uncomfortable positions, making the convenient bottle and handle the standard for the antifreeze industry.

The year this bottle was designed, it was given the Package of the Year Award from the Industrial Designers Society of America, and people kept remarking, "Why wasn't this done before?"

What makes your life easier?

Do consumers really want to pay extra money for certain conveniences? There are opportunities for you to provide conveniences on packaging without additional costs.

Coke and Pepsi now have refrigerator packs which, once in place in the refrigerator with the flap removed, self-dispense the soda cans. This eliminates the need to open up the pack and stick each of the 12 cans into the refrigerator as single items. Great idea and no additional cost.

The convenience of the slide-in package helps your store sell more cans and simplifies things for the consumer

Get 12 for the price of 10? It's important that you put a quantity of items together in a package not only to reflect an attractive price, but also to reflect what the customer actually needs.

If you're selling fruit bars, should there be six, eight, ten or twelve? Polling a group of consumers can be helpful for this rather than depending on putting out a package that just gives the impression of the right price.

section three

connecting
the package with
the consumer

Chapter 9
Shaping up

"If it ain't broke, don't fix it." A popular expression, especially in the packaging industry, where some think it's expensive to make any kind of change.

But it's really more expensive *not* to make the change, since the consumer is always on the lookout for something new.

This is especially true with the structure and shape of a package and how that package can influence convenience and recognition. That physical shape is important, and don't think shoppers are not aware of it.

We mentioned earlier how influential package shapes could be from responses to plain black silhouettes of bottles of ten different liquor brands. Since practically 100% of the people at a seminar recognized all ten liquor brands just by those silhouettes, this was either a heavy drinking crowd, or bottle shapes aren't easily forgotten. It's undoubtedly the latter.

Needs improvement

Advertising messages often change, but in some industries, packages – especially their shapes – hardly ever change.

Research firms have found that today's shoppers want portability, resealability, improvements in grip and in opening and closing the package. Big opportunities in the future for packaging are with the innovation of the package structure.

Older consumers may still respond well to the earlier package, but young consumers need something different, and all industries have to realize this. Even Absolut Vodka, who built its campaign around its package shape, considers changes every so often to accommodate a younger audience who often looks for a dynamic new package.

We're seeing advertising messages relying more than ever on the product package. Take any popular magazine, and a large proportion of the ads will show a prominent package shape as will TV commercials.

Your big opportunity for improving package structures is the inte-

The ad cannot just address the product; it must show the package so the consumer can find it at retail

gration of the package with the display. It really started with the encouragement of the big retailers, especially in big-growth, point-of-purchase categories.

We developed unique and visionary display structures several years ago for Eveready, which integrated with new package structures, making innovative point-of-sale units. Duracell, Eveready and others compete strongly to win retailer space through structures that are eye-catching, functional and easy to shop.

The visionary pantyhose

How do you find new places and new ways to market products that people are used to finding in department and retail stores?

Women's hosiery had been around in many forms, but Hanes, in North Carolina, pioneered the introduction of pantyhose in the mid-1960s. Its position was well established in department and retail stores, but the company had a vision that its products should expand into other sales channels. In the late 1960s it saw the growing food/drug/mass-merchandising stores and realized that this could be new territory for its products. The vision was if women shop there, let's go there.

The company introduced a whole new merchandising and packaging concept called L'eggs in 1970. Its pantyhose was packaged in an inno-vative plastic egg-shape integrated within a display, which itself was the shape of a giant egg. This package and merchandising unit cap-tured the imagination of retailers and consumers.

Carrying through the vision, the integrated packages and their displays were placed in supermarkets, drugstores, convenience stores, mom-and-pop shops and even laundromats. Every opportunity was used to get it in

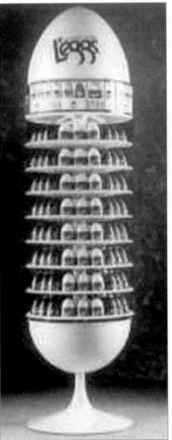

front of the consumer. The L'eggs concept carried through as a total brand, using the name, package shape, graphics and display to entice the consumer.

Although the egg-shaped plastic package originally introduced was somewhat gimmicky, it worked to build the brand. With the environmental concerns of the 1990s, L'eggs replaced the plastic egg with a cardboard package using less material and made from recycled paper.

The L'eggs packaging and marketing story is considered a classic. Since so much of the concept revolved around the package and the display, you can just imagine how many designers, marketers and manufacturers claimed that they had had the original vision for this package. Our firm has interviewed many designers through the years, and more of them than is possible have indicated that they had the idea!

The visionary L'eggs brand combined good merchandising with a unity of name, package and display

No matter who actually had the original vision for L'eggs, it worked to build a brand that is memorable. L'eggs remains the best-selling pantyhose in America.

Hold your breath

Functional benefits can produce packages that will eventually become real winners. We've seen how lifestyle and usage occasions have changed for the consumer. Because of this, portable products are becoming mainstream.

A great example of a vision for portability is the new Listerine Breath Strip packages. These breath strips have moved into the empire originally established by breath mints. Small portable packages like this with a new type of product takes substantial engineering.

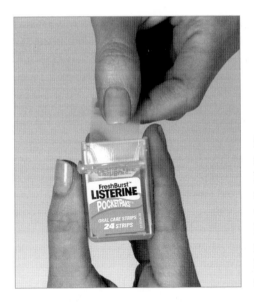

Many users abandoned TicTacs when they found a package that was smaller and much quieter when walking around

But when marketers realized that a fresh new approach was needed for breath freshening and discreet portability, they answered that need with brilliant vision and engineering, by working from both the inside out and the outside in.

The visionary mouthwash

Let's talk some more about Listerine. Everyone knows the brand – in fact, it's over 100 years old, and is the largest selling brand of mouthwash. For 100 years, Listerine had been packaged in a glass bottle covered with a corrugated sleeve and then overwrapped with paper labels.

Why change the number one brand? Well, for several reasons. Listerine users were getting older, and younger users were moving to the more flavorful mouthwashes.

The overpackaging was an issue, with both cost and environmental concerns.

Any change would be a major challenge, since the Listerine brand was widely recognized through its packaging. In fact, on the retail shelf, Listerine products could be spotted 50 feet away, since they were the only products packaged in this fashion – the same packaging it had used for 100 years.

But a change was needed. In order to develop a new bottle with the brand character of the Listerine product, our step-by-step process included:

1. A complete audit and analysis of the category, looking at store shelves all over the country and looking at our client's manufacturing methods. Although looking old-fashioned, Listerine was by

far the easiest product to spot on the shelf. At our client's plants, we noticed that the glass bottle lines were slow and produced breakage along the way.

2. Research was conducted to understand consumer attitudes toward Listerine. As you can imagine, Listerine was considered a very strong product. Not very good tasting, since the flavors had not been introduced at the time, but it worked especially well. As some people would say, "If it hurts, it's killing the germs."

3. A brand platform was developed for Listerine, which included the vision of a product line that would be considered strong, supreme and effective in all attributes of mouth washing.

4. A range of bottle shapes was developed to generate the feeling of the brand itself – efficacious, hard working and strong. We also wanted some of the bottle shapes to suggest the recognized "barbell" shape that Listerine used for so many years.

Many models were made from the range of bottle shapes. The models were tested, and the winning shape was a contemporary version of the barbell shape. The angular shoulders and sharp corners of this new shape suggest the strength and efficacy of the brand.

To top it off, a dosage-sized cap is used.

The graphic format and the label color suggest a relationship to the well-known Listerine brand colors.

The new bottle, together with newly developed flavors, was proportioned for eight different product sizes, and now appeals to a younger audience without alienating older users. This visionary bottle maintains the classic brand equity built up over 100 plus years.

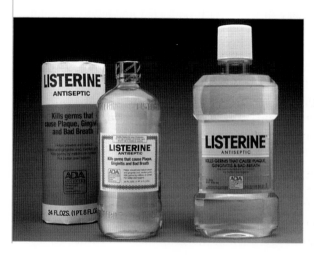

It takes vision to risk a major change to the package of the leading brand. The risk was minimized by market testing the new bottle in Canada for several years before bringing it to the US market

The visionary Listerine package has many imitators within the store brands, but still is recognizable quickly with its consistent sharp angles and graphic colors.

The strength of the brand has enabled a wide variety of line extensions, including the new, very successful Listerine pocket packs described previously. These pocket-sized oral care strips are overtaking the breath mint tablet category and bring another visionary package to this wonderful brand.

The visionary ketchup shape

Ketchup as art? Sounds strange, but how do you create interest in a staple product that's been around for many years and is used the same way on food products year after year?

Well, just about everybody has Heinz Ketchup in their home, and the company felt it was time to make some bold moves in order to expand ketchup usage.

Heinz had the vision that ketchup could be used not only as a condiment for making products taste better, but perhaps also for having fun. They had noticed kids playing with ketchup at the table, sometimes spreading it on food in unusual ways, and even making patterns with the bright red color.

Together with Interbrand, Heinz teamed up to reinvent everything about the Heinz ketchup brand except the taste itself. If kids enjoyed products that called for ketchup, such as hamburgers and fries, why not combine that with the fun that kids could have in putting patterns on these products? Or using the ketchup just to have fun in general? This was a new brand attitude and a product targeted for the kids.

A visionary package was needed which could become a vessel for holding the ketchup as well as for spreading it in a fun manner.

How did we implement this vision? We watched kids at play, and designers observed children of various ages eating with ketchup and playing with ketchup. Kids were watched playing with other items as well, in order to get the idea of how they interacted with colors, painting and making creative patterns. The vision was for this package to appeal to the kids.

A wide variety of concepts were developed, with many bottle shapes and many delivery systems or spouts. Some ideas included pumping, pressing, pushing, squeezing and pouring out the product.

After much experimentation, a shape was established. This rounded shape, that fitted easily and comfortably into a child's hand and even looked childlike, was combined with a spout that would make the ketchup pour in a thin stream. The thin stream could make interesting patterns while the bottle was squeezed.

Matching the brand concept and the attitude, a new product name was developed – EZ Squirt.

Squirting red only? Never. Green, and then purple ketchup was introduced.

This uniquely functional bottle, with appropriate and communicative graphics, was an immediate hit. It has become one of the success stories in the food industry, and it was created by people with a vision who saw the writing on the food.

The package communicates that it's strictly for kids, and the Heinz brand quality is conveyed through the white keystone shape

The visionary tuna package

The tuna fish category is a tough place for a brand to swim. Through the years tuna has been sold in small cans, all with similar images, identities and brand promises. Because of the size of the can and the small space for the label to convey so much information, the category is difficult to shop. Consumers are confused about the brands and varieties of tuna. Is it white or albacore? Is it water or oil? Or what is it? How do we make it easier to shop for tuna fish and provide some added benefit?

We spent much time shopping with consumers in the tuna fish section in order to understand their thought processes, determine how they approached the tuna shelf, and discover what messages to emphasize on the package.

Our client, Starkist, had used the spokefish, Charlie, for many years. The challenge was to create a differentiated look for Starkist while capitalizing on this beloved spokefish, Charlie.

So we needed to establish the correct communication pathway.

Tuna fish was taken out of the can and packaged in a pouch – a Starkist first! The foil pouches took up no more space than the can, but enabled a large front panel to be used for product photography and product information. A blue background was used for all the graphics and labels, to further differentiate the brand from anything else on the shelf. And Charlie became the star of the package.

The use of foil and the new pouches and blue color set the tone and image for the Starkist brand, giving it a prem-ium look and a new vision for the future.

Starkist tuna made the biggest splash of all the brands by being the first out there in a pouch

The visionary paint

It's a struggle to use paint, and setting up is often the worst part. Espe-cially since the standard metal can has remained essentially unchanged for over 100 years. While you're painting, the excess paint pools around the lip of the can and, when finished, closing the can is another mess.

Overcoming 100 years of inertia in paint packaging requires a visionary approach. Nottingham-Spirk Design Associates and Sherwin-Williams combined to develop a revolutionary paint pack-aging system.

It seemed as though everyone in the paint trade could think only of reasons why a plastic, portable container would not work. Obstacles involved fitting existing shakers in retail stores, stacking in ware-houses, filling-line equipment and perceived drop-testing problems.

With vision, the development teams could sense the potential con-sumer demand for the new packaging system, if they could overcome the obstacles in the creative process.

So the designers studied the entire life cycle of a paint can, traveling from the can manufacturer to the filling lines, through distribution centers to the retail stores, through the in-store tinting process to the consumer's home, to usage, storage, reusage and the eventual disposal or recycling.

Round container shapes, square container shapes, and a wide variety of product features were conceptualized.

It was also noticed that more women than men today are making the key decisions about paint, and are more demanding in its ease of use and its neatness for do-it-yourself projects.

To reinvent the package, it was also necessary to stay within the standard gallon can footprint, so that warehouses and retailers would not need to change their shelves.

Of course, the opening in the package had to be large enough to fit a standard 4-inch brush and still accommodate a pourable feature. Many different container concepts were tested against the standard paint can, and the plastic pourable can had an overwhelmingly positive reaction.

The reinvented container did require distribution changes. Store shaking machines had to be outfitted with square inserts for the new container, paint filling lines needed to be converted for the square plastic containers, and warehouse pallets needed to be changed.

This visionary package has provided unexpected benefits. For example, it's improved the store tinting process, since opening and closing the container to add the tint is faster. The container requires less shelf space, and provides for 14 package facings rather than the current 13 facings for metal cans. And a flat front package face gives a strong visual label area.

The container was launched using the Sherwin-Williams Dutch Boy brand, and it appears that sales are exceeding expectations. In fact, the easy-to-use containers are exceeding consumers' expectations.

This reinvention of the gallon paint can finally makes it easy to open, pour and close.

It's about time.

Now you realize why you never liked the metal can

Visionary vending

Perhaps you've been to Japan, where buying food and drinks in many of the restaurants can cost you a fortune. Just a cup of coffee can often set you back quite a few dollars.

But the Japanese have found a way to save money on their precious coffee through their vending machines.

Canned coffee is actually a Japanese invention, and is most popular among men in their twenties and thirties who don't brew their own blends and don't care whether or not their coffee comes in porcelain cups.

For over 25 years, canned coffee in vending machines has been a popular drink in Japan, commanding the largest single share of Japan's domestic soft drinks market. The cans from vending machines account for about 20% of the country's coffee consumption. And Japan has the highest per capita number of vending machines in the world, capable of selling both warm and cold beverages. With the capability of drinking canned coffee hot or cold, the vending machines are active throughout the year.

And since fewer female workers are willing to serve tea to their male colleagues than they had been in Japan's past, vending machines with canned coffee have also found a very respected place in many offices.

There are now more than a hundred different canned coffee products on the market, some quite exotic. Producing the coffee requires special technology, especially since it's often kept at very high temperatures and in a vending machine for months. The coffee usually comes in steel cans, which are stronger than aluminum containers.

Originally, it was visionary to introduce hot and cold coffee in cans in vending machines. But the new vision is in the can *shapes* that are currently surfacing in the vending machines, to differentiate new brands and new blends from the others. JTI Beverages of Japan, the company that is emphasizing the new can shapes, has introduced some unusual shapes in a world of cans noted for their conventional shape. One of their products, J T Roots Real Blend, is in a can shaped like an hourglass.

But its visionary package is more functionally distinctive. The can for the upscale Roots Rcafe Cafelatte has an insulation strip around the body of the can, making it easier to hold when the can is hot. The insulation significantly lowers the can's temperature, but only on the outside. Additionally, the can has a narrower bottom, which reduces the time it needs to be heated inside the vending machine from 20 minutes to only one and a half minutes.

For coffee vending machines in Japan, the visionary Roots packages are functional as well as distinctive

With this visionary package, the Japanese will be able to enjoy their coffee steaming hot, and with improved flavor and texture. This new can introduction into the ever popular vending machines is already proving its comfort and popularity with the consumer. It's the shape of things to come.

Adding a handle

Did you know that bottles with handles cost more to manufacture than bottles without handles? There's always a debate as to whether the handle will encourage the consumer to buy and use the product, or can the producers reduce costs by not adding the handle?

Before the popularity of plastic bottles, all liquids were packaged in glass. Some large glass bottles had one-finger handles to allow for a grip, especially when pouring. Over 40 years ago, Clorox bleach was an example of a one-finger grip glass bottle known by its shape. When it converted to plastic, the Clorox marketers originally insisted that the plastic bottle should look exactly like the glass bottle, one-finger grip and all. This was typical of many companies converting from glass to plastic – they were afraid of changing the shape, even though plastic can be molded into a million more shapes than glass. And handles can be all sizes and much more comfortable than one-finger grips.

This attitude of retaining the old glass shape lasted for several years until brand marketers finally realized that they could take advantage of unique and innovative plastic shapes for their bottles. New gripping devices have emerged, but brand marketers are still faced with the dilemma: "Is it worth the extra money for a fancy handle?"

In most cases, yes. But understanding the shape and structure and its relevance to the brand is important.

Before Murphy's Oil Soap sold its business to Colgate-Palmolive, it was started and owned by the Murphy family of Cleveland, Ohio. The Murphy brothers, Paul and Ray, were looking for a way to grow their brand. The oil soap was a marvelous product for cleaning wood cabinets, wood floors and most wood household items, but their packaging was lackluster, with cylindrical bottles which were not as convenient as other cleaning products.

The largest selling product was the 28 oz. size, and we noticed immediately that the cylindrical bottle was clumsy and difficult to hold, turn and pour. We strongly recommended molding a handle into a proprietary bottle configuration. The Murphy brothers resisted our idea at first, thinking that the cost of changing their molds would be overwhelming. We were able to convince them that it would be worth it in the long run because of the convenience, and the original outlay for changing the mold would be amortized as time went on.

Without any changes in advertising or promotion, the new bottle was introduced with the handle. Sales moved so quickly, the Murphy's could hardly keep up with demand. This proved to us, without a doubt, that adding a handle to this product increased consumer satisfaction and sales.

The convenient bottle and the proprietary shapes that were the result of that design program on their other bottle sizes enabled Murphy to build its brand.

Although we changed the package shape, we were able to maintain what was referred to as a "humble" simplicity in the style of the package. The Murphy family was happy that they had relented and allowed this new handle to be incorporated into what still remained the Murphy shape identity of simple and humble lines.

And what a difference a handle can make! People no longer had any fear of getting the product on their hands. The handle made the bottle easier to pour and sales increased to a point where originally it was hard to keep up with production.

No advertising promotion or coupon deals could ever have increased the sales the way the handle and its convenience and shape moved Murphy's Oil Soap.

After the handle was molded into the bottle, the Murphy family was amazed at the brand's immediate sales growth

Quality counts

Good quality costs more. It often does, but like architects designing houses, good designers can work well with bottle suppliers to get the best quality in custom package shapes at affordable prices. And good production quality is especially important in certain categories such as cosmetics, where the physical shape of the package often makes the sale.

"If a bottle looks rounded, and soft and embryonic, the consumer will be disappointed if it feels rough when she picks it up," according to Marc Rosen, the well-known cosmetic designer. If a perfume bottle has sharp edges, which can sometimes look sexy, and she touches it but the edges really are rounded because the production was not especially good, the consumer will be disappointed and maybe not want to smell it or buy it.

Every point we make can be influential, especially when it comes to shaping the package.

Chapter 10
The package in the retail store

In ten years you won't be watching TV commercials. We've all read articles about how TIVO and other replay boxes are going to enable us to watch the TV programs when we want and also skip the TV commercials when we want.

Perhaps it is an exaggeration to say that no longer will you watch commercials at all, but undoubtedly other ways of promoting products are going to be more important than ever.

The store is becoming a more important marketing medium, and packaging and design within the store is the big opportunity. This means a more dynamic store environment, since people will be making their decisions more in-store and at-shelf than ever before. In fact, it's estimated that over 70% of purchasing decisions are made in the store.

The big brands are worried because they are losing their traditional connections to the consumer. According to a recent *Fortune* magazine article, it only took three TV commercials in 1995 to reach 80% of 18- to 49-year-old women. Five years later, it took 97 commercials to reach the same group. So your package has to work harder.

If you take summer vacations, you must have noticed how each year the retail landscape substantially changes. A new Wal-Mart is in the neighborhood, or Lowe's, or Costco or whatever. There are eight to ten dominant retailers who have increasing power and realize that they can affect the traditional media and traditional methods of marketing.

You are probably attracted to people because of their personality and their looks. You are probably also attracted to retailers because of their personality or their looks.

Impulse decisions at the shelf demand that the package design be compelling, brilliant and arresting. The package needs to become more of a promotional and advertising tool than it had been in the past, and stores have to think of ways to win the customers without relying solely on price. So if you used to just sell the brands that were made and marketed by the big manufacturers, you now need to behave like a marketer.

The big change

Do you remember when U.S. supermarkets changed their own brands quality?

This big quality change showed up in two stages during the 1980s. First there was the range of generic or "no-name" brands sold in the supermarkets. The idea was to make the shopper feel as though she was saving a fortune through brands that didn't advertise and used "aggressively bland" one- or two-color labels. Some chains took full-page ads to announce these economy products.

Since our branding firm specialized in package design, a reporter for the *New York Times* asked us if this meant the end of our profession. "No way" was the answer. "In no time the generic brands will be competing heavily against one another and will need package design more than ever to differentiate their brand attributes."

But even before generic brands had the time to catch on, U.S. supermarkets changed their own brands' quality. Loblaws in Canada was probably the pioneer in North America, after seeing the successes of Marks & Spencer and Sainsbury's in the U.K., Aldi in Germany and Carrefour in France. Loblaws started with an extraordinary tasting

chocolate chip cookie called "The Decadent", and moved many other products into rich ingredients foods with quality higher than national brands. This was a visionary approach from Loblaws' CEO Dave Nichol. The package designs imbued quality and, in many instances, looked better than the national brands.

The Decadent package design was an early and successful store brand attempt to outdo the appetite appeal of the national brands

Shortly thereafter, in 1988, our office was pitching brand identification to a Safeway vice-president at his office in Walnut Creek, California. The Safeway VP had a memo stating that Safeway should start to develop its private label packaging with a line that would be on par, and even superior to, national brands. The memo also stated that the package design must communicate the quality of these Safeway products. Since Safeway was one of the largest grocery chains in the world, we realized that this approach to private labels in the U.S. would be the wave of the future. And it was.

Why does private label work?

Package designers get calls from the management of retail chains and some manufacturers requesting private label design. Some actually request that their package be designed to look exactly like the leading national brand – probably with the thinking that the customer would mistake one for the other. Our firm always refused to do this, but I'm sure the chains found someone to do it for them since you can still find many store brands that are hard to differentiate from the well-known brands. But imitation no longer works.

The national brands always had the money and took advantage of innovating, and the store brands just copied. But store brands can't work this way anymore.

Many stores can now pull shoppers in because of their own branded products, a trend that opens up new opportunities in retailing, especially in supermarkets. Clothing stores like The Limited or Gap have always pulled in customers with their branded products, but this is comparatively new for supermarkets, which have developed particular products to a level that the consumer will make special trips to those particular stores for their private branded products.

The Publix supermarket management have vision. Everyone in Florida shops at Publix stores, known for their good customer service. Publix actually started out as a dairy company and has one of the largest dairies in the country. It applied its dairy knowledge to develop a marvelous-tasting premium ice cream. After seeing some publicity about the package redesign work we did for Breyers Ice Cream, Publix contacted us to redesign its premium and standard ice cream packages. Why? Because Publix knows that it can bring extra traffic into its stores with the best ice cream product and the best ice cream packaging while maintaining and growing its strong ice cream franchise.

The strongest private label brands are in the U.K. A number of years ago, Tesco, the major U.K. retailing chain, had a vision to take in a broad range of customers and satisfy those customers all in the same store. John Clarke of Tesco referred to this vision as "the spectrum of shopping experience". Tesco would have products focused on the low, and top end. For some of its food, the customer may be adventurous and try new things and pay a bit more. The same customer may think of other products as a commodity and want to reduce their bill by buying a cheaper brand. Tesco shoppers say that they do not want to be tied down and want to shop as they like, and Tesco caters to people wanting that experience.

Tesco's philosophy focuses on making sure its customers have a full shopping experience when they go into the store, understanding that its product will always be there along with a wide range of products. The Tesco personality is different from its competitor, Sainsbury, who caters more to the higher end. The whole personality of Tesco brings people into its store, with signage that's easy for the customer to follow and background colors on the walls to make it intuitive in terms of where their product will be.

As a retailer, your success is predicated on understanding your own personality. The store brand also has to function and enhance the store's personality.

The retailer's personality needs to interact seamlessly with the shopper. It needs a point of difference through its private label brand, and this private label brand is its corporate identity. Before coming out with this new private label brand, the retailer must understand who it is, what makes it tick, what consumers think of it, and then develop the brand that supports those issues.

Retailers should forget about the big worry of getting their product on par with the national brands. Just finding the right sources of supply enables them to produce a product with quality equivalent to 99% of the national brands.

Some of the stores today have as many as 3,000 or 4,000 store brand products on their shelves. For example, Kroger, the giant U.S. grocery chain, makes 4,300 food and drink items in the 41 factories it owns and operates. The trick is embedding these products into its category planning structure and getting its brands to walk, talk, act and feel like brands that appeal to the consumer. Visionary package design can do this.

Where in the world is private label?

First let's talk numbers. In the U.S. private label brands account for 22–23% of store merchandise. In neighboring Canada, the Loblaws chain has about 40% of its products in private label. We already spoke about the U.K.'s dominance in private label. In fact, Marks & Spencer is 100% store brands, and Sainsbury about 60%.

Asia is in a new stage of development where retailers have 7–10% of their products as private label products. Because they are at a different stage with store brands, developing shopping behaviors will be an important opportunity. They also have a big opportunity to bring in new products, promote them frequently and make sure that the store brands are married to the basics of their store environments. Along with good package design and good signage, this could be a recipe for success.

The destination brand

What do Michael Graves and Philippe Starck have in common? The answer – they are well-known designers who are invading mass merchants. And these retailers borrow the equity of these well-known designers for their private label branding programs.

Attitudes toward private label branding are really different across the world. But all retailers see it as a way to enhance margin and differentiate themselves. Destination brands – some may call them fashion brands – have become part of many retailers' portfolios and are a big opportunity in the future.

Target Stores built a style and fashion sensibility unlike any other store, borrowing equity from, or in some cases creating equity, from people who many consumers were not familiar with. Unless you were from architectural or design circles, there is little chance that you knew of Michael Graves, so Target brought the Michael Graves name vividly into the store, promoted it highly, and you now have lines of Michael Graves' housewares with a unique kind of fashion sensibility and great packaging. Target did it within a bright, airy, cheerful environment that distinguishes itself versus Wal-Mart and Kmart. This is the Target personality, with the products, the packaging and the fashion sensibility blending as an example of its own branding. An estimated 50% of Target's packages are designed exclusively for its stores.

The Michael Graves designed products and the packages make a unique and fashionable display, enhancing the Target image

You can see Target commercials on TV where there's hardly a mention of the Target name, but their distinctive way of advertising with a flair, sometimes even before you see the Target logo in the commercial, tells you who it is. The big opportunity is matching up the private label strategy with the retail and promotional strategy.

Imagine the strength of retailers, when you realize that a company like Wal-Mart has over 20 brands within its stores that are proprietary. And Wal-Mart is going to put everything it has behind its own brands versus national brands.

Some stores have dressed themselves up through their packaging. H.E. Butt in Texas takes a leadership position in delivering packages with unique graphics and structures.

Meijer, from Grand Rapids, Michigan, has been successful in creating excitement within its store space with destination brands. By developing names for product lines, such as "Baby Beginnings" for baby clothing, its own products converge with its brand strategy.

Shoppers can find a wide range of baby clothing by looking for the Baby Beginnings tags, one of the many destination brands at Meijer

People go to Trader Joe's for the purpose of buying their store brands. Trader Joe's has been visionary in its whole marketing approach, and with the cachet conveyed by its product lines and packages. It looked at chains like Safeway, Kroger and ShopRite, and felt it could use its Trader Joe personality to strongly compete.

Made from Trader Joe's own vineyards, its Charles Shaw wine brings in customers with its amazingly low price, good taste and pleasant package. If stores like Trader Joe's can bring in customers with its own brand and special packaging, why shouldn't this trend open up new opportunities in supermarket retailing? Supermarket retailers are selling a wide range of products, including computers, housewares, air conditioners and fine wines. The big challenge is with the packaging. Remember, the marketplace is free and exciting and whoever comes up with the best packaging wins.

A main attraction at Trader Joe's is the packaging. Almost exclusively store brands, the exotic package designs create strong brand loyalty among the shoppers

The private label cookie-cutter

Should toilet paper, ketchup and shampoo have the same packaging? There are many schools of thought in private labels, with designers and marketers having different opinions. However, the design approach seems to come down to three strategies.

The first strategy is a similar brand identification across all the products, looking somewhat institutionally driven. This is the *rigid* guideline strategy, which makes the labels on the toilet paper, ketchup and shampoo look very similar. This can work when the retailer wants to make a statement that is more about the brand than the product. For example, "My platinum-colored packaging shows that it is the most premium product," or "My red packaging shows that it is the best budget product."

The second strategy is more flexible, or category-specific, similar to

Wegmans, with a loyal customer base, uses a rigid guideline strategy – the store logo is over the product name which is always in a black panel. Their store brands are easy to spot and compare with prices of national brands

the Kroger or A&P approach. The retailer addresses the package design towards certain dynamics within the category. The toilet paper package may emphasize softness through its background; the ketchup may emphasize appetite appeal, showing its source of tomatoes; and the shampoo may emphasize scent. With this flexible strategy, the retailer uses tiers of quality with a different name for each tier. The label designs suggest the quality of each tier.

Kroger's package designs are flexible and category-specific in order to compete head-on for quality with national brands. Its three tiers – good, better, best – are represented by FMV (For Maximum Value), Kroger, and Private Selection

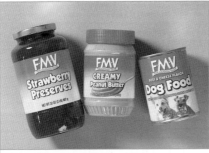

Guidelines are flexible for A&P packages. The designs reflect the nature of the product category. The America's Choice brand has two levels of branding to reflect their shopper. One level uses the America's Choice brand directly with the product name. The other level uses a category-related sub-brand name endorsed by the America's Choice brand. Master Choice is their premium store brand for specialty foods

These are different ways to segment store brand programs. The retailer must understand their own personality, along with understanding in which categories they can emphasize their strength.

Retailers need branding discipline and this is the big challenge. Remember, the retailers' background in the past was in procurement, not in developing their own brands. Today's designer can show them that there is power in developing these store brands through innovative design, structure and new products. This is the way to compete against the Wal-Marts and against the store across the street – create something that is truly ownable, innovative and emotionally relevant.

We've already got consumers to understand that private label brands are usually cheaper. They also believe that most of the large chains have them. And you know that a trusted store brand can differentiate a chain from its competitors.

The premium folly

"You ain't what you ain't." Retailers have to know who they are. Pathmark, a large northeastern U.S. chain has a totally different personality from Harris Teeter, a midwestern U.S. chain. Both chains have different consumers, with Pathmark having a more price-dependent consumer. Pathmark has a successful image in the marketplace, offering low prices and good value.

But about eight years ago, Pathmark looked into a premium branding trend, and decided to offer a line of products called Pathmark Preferred. They figured that premium branding is a trend in the industry, which will bring them some new customers. The Pathmark Preferred packages were designed to look glamourous. But the design did not relate to the Pathmark customer. Beautiful and compelling from a visual standpoint, it was actually overdesigned, since it did not fit the heart of who that particular retailer, or their consumer, really was. The brand needs to conform with the essence of the store, in this case, Pathmark, and what it is really known for in the marketplace.

The cereal

You can put four boxes of cereal in your shopping cart and possibly faint at the checkout counter when you see the price rung up. There are jokes about people mortgaging their houses in order to buy cereal. And believe it or not, cereal companies complain that they are losing money, even though you may pay $4–$5 for the single box of cereal that used to cost much less.

The cereal companies claim that the cost of the ingredients has risen substantially. Some consumers believe the cereal companies have increased their advertising spending and their package expenditures. We know that the printing and paperboard on packages have rising costs, adding to manufacturers' worries.

But the real enemy of the national cereal brands are the store brands. Without the need for the advertising that the national brands have, and without flamboyant packaging, their prices for similar products would be considerably lower.

We all remember when there were many fewer cereals, and each brand displayed its own line of products together on the retail shelf. Not any more. The raisin bran products are all displayed together – national brands together with store brands. Cornflakes are all displayed together – national brands together with store brands, and so on.

This means that you no longer have the big Kellogg's, or General Mills billboards where their products were together. It also means that you might see Kellogg's Cornflakes for $4 butting up with store brand cornflakes for $2.

If national brands are unable to lower prices, there is one way for them to overcome the consumer's thought of buying the cheapest cornflakes or whatever, and that is to capitalize on the brand itself. Companies like Kellogg's and General Mills should pump up the name or the brand identification of Kellogg's or General Mills so that a brand loyal public will stick with them. This doesn't only apply to cereal, but to any product category of national brand that is endangered by the store brands.

All the cereals are emphasizing their products with gigantic photos and catchy names and seem not to emphasize their fine national brand name. This is a mistake. Since these national brands are spending considerable money on advertising and promotion, instead they need to emphasize that brand name on the cereal package itself, right across the board.

An example of today's store brands, Albertson's Cornflakes has a much lower price than Kellogg's, along with appealing product photography. To compete effectively, the national brand, or Kellogg's, should flaunt its brand name on the package

It can't look better than it tastes

You can't make a first impression a second time.

Tens of thousands of products and packages are staring at you as you march past the shelves of your supermarket. Whether it's a national brand or a private label brand, the package makes the first impression, and it can't overpromise.

National brands have known for years that consumers won't repurchase their products if what's inside the package doesn't meet the expectation from the outside of the package. Private label brands, less experienced with package experimentation, are learning how to attract their customer by the packaging. But the product inside must be equal or better than the way it looks on the package, or the second sale will not be made.

We're not suggesting that the package photograph should show the product in its dreariest look. No one would go to an interview wearing his or her dreariest suit. However, consumer frustration from a product that doesn't meet up with expectations from the package reflects on both the manufacturer and the retailer.

Weight Watchers has a line of food products, which capitalize on their well-known diet plan brand. At one time, they explored packaging with luscious presentations of meals. The package designs were brought into consumer research and were shot down. The food photos were so fancy that people couldn't believe that the food inside the packages could be that good. Weight Watchers took the packages out of the running.

Give them what they need

"There is a great new opportunity for package goods companies. Stop worrying about the consumer only – start looking closely at the retailer and figure out ways to meet the retailers' needs", says Bob Swientek, editor of *Brand Packaging* magazine.

You need to understand the retailer's merchandising methods and its packaging. You need to look at whether there should be a special size that the retailer might want to use to differentiate itself from the others. Perhaps a special promotional idea can tie it into a larger store promotion, and you can provide a complete merchandising plan, which should be arranged to suit the retailer's needs. Get a Target, a Wal-Mart or a Kroger for an exclusive arrangement, which helps them and obviously helps you, because it puts you more in touch with and aligned with what the retailer is thinking. It helps you to understand their future thinking and their future planning.

Any time a brand marketer can understand what the retailer really wants, and where the retailer is headed, gives them an advantage over its competitor. These are opportunities that designers and marketers should run after.

We know that impulse purchases are going to rise, so obviously packaging and merchandising should become more important in a brand's marketing campaign.

The efficiency expert

Costs are rising out of control. As a goal, packagers need to stay simple. According to Bob Anderson, vice-president manager of Wal-Mart's private branded products, Wal-Mart helps to keep their manufacturers' costs down by using standard printed shipping cases and, through LaserJet, changing certain words to denote particular products. You can compare this to "just in time" packaging, or "mass customization". Since large inventories may not be necessary for each flavor, the inventory is made for the product line itself. The flavors are added as needed. The manufacturer saves money with the cases, and the store doesn't stack the products that are not immediately sold.

Saving time for the store clerk saves money for Wal-Mart, and displaying packages in their shipper saves unloading time

The global retailer has different concerns from the regional retailer. Since temperatures in the U.S. vary from the cold north to the Arizona desert, retailers can experiment with packaging and products there more easily than in other areas. But the packages must be designed so that everything is handled efficiently on its travel to the retailer. Fresh produce must stay refrigerated. Ice cream must be able to be put away quickly and food products need to keep within their required temperatures.

The visionary shipping cases

We often forget about the shipping case itself when we talk about the package, but many products, especially houseware and cookware, are usually sold in their shipping case. And for many years those shipping cases were dreary and even hindered the sale of attractive houseware or cookware items.

One of the most well-known lines of cookware is sold under the Farberware brand name. There are actually hundreds of items in the Farberware cookware line. This line is divided into many product categories, including stainless steel cookware, serving plates, utensils and many other lines. At one time, each of its categories used unrelated packaging components and packaging graphics, which became especially apparent when parts of the line were displayed separately in various departments throughout retail stores. Some of the Farberware products were in lightweight packages with several colors, others were in heavyweight packages in one or two colors, and the brand was not coordinated. It didn't look like one single brand.

Our vision was to develop one overall look for Farberware, along with packaging that was consistent throughout the various categories and products in the Farberware brand. We wanted to give them beautiful packages.

But would you want to buy a beautiful package and then have it stuffed into a shipping carton to protect it? And these products really needed protection since they contained a wide range of heavyweight cookware. Corrugated board was selected as the structural material for the packages.

But corrugated usually meant the use of a somewhat crude printing process called flexography which was especially expensive when used in many colors. Yet our vision was to show these products at their best. We wanted to use the finest photography to emphasize the shape and quality of the products. We wanted to enhance the feeling of the products with props, such as vegetables and fruits.

A unique and visionary approach was taken for these corrugated shippers. Large paper sheets were offset printed in four colors, achieving a quality photographic image for the product photography and the graphics. The sheets were cut into large labels with grey borders to match the grey color that was applied to the corrugated board. This gave the impression of a seamless package of great quality to show off the products.

These visionary new packages unified the Farberware line and brought new recognition to its brand name and its diversified products. Farberware was the first brand in the category to use this type of packaging and printing technique, and gained a strong advantage over all its competition. It has kept its lead in the marketplace.

This visionary approach to corrugated cartons revolutionized the combination of product protection and quality display

Helping the shopper

Let's compare two types of stores. We'll call them Store A and Store B. Store A invests more in packaging and its supply chain, and invests in its packaging design to apply to different types of customers.

Store B invests in its packaging, but also in its replenishment systems, and in its content proposition to customers. Store B wants to know more about the customer as they walk in. It wants to know what the customer wants and to have the correct ranges of products in a place where customers can find them. According to John Clarke, this is the route taken by Tesco, the visionary U.K. chain.

But isn't packaging also important? Mr. Clarke relates packaging to its product placement within its stores and the ease of shopping. Tesco identifies the brands it sells in its stores on three levels: Tesco Value, brand goods and Tesco's Finest. This strategy is interesting since Tesco uses their two own labels to flank the manufacturers' brands.

Tesco Value, at the low end of the price spectrum, targets the thrifty shopper. It's a simple white package with blue vertical lines. At the opposite end of the spectrum, Tesco's Finest appeals to shoppers whose priority is higher quality. The package is more substantial, with silver printing and an upscale appearance. The manufacturer's brand is caught in the middle, having to compete against both ends of the spectrum. However, the manufacturers' brands are also a huge and growing part of Tesco's market.

As John Clarke said: "If you want to impress your girlfriend or have some friends over, you buy the Tesco Finest in its silver packaging."

So Tesco has some unique packaging, but the company also focuses on replenishment and customer insight. They are always looking at the product solution using vehicles such as their club card, getting data into its warehouses, analyzing that data, and trying to figure out the correct range of products to have in a particular store within a particular location.

Tesco's plan-o-grams are specific to its individual stores, not general plan-o-grams that are used across the board. By getting to know more about its customers within specific stores, it is able to set its range of products and the types of packages the customers really want. This is what John Clarke refers to as "range replenishment and consumer insight".

What's in the future for Tesco? A broader range of offering, which can be focused on aspects of lifestyle. An example of this would be expanding the stores' organic section and the packages within the section. Also Tesco is historically a food store, but within a few years, its non-food business will equal the size of its food business.

Tesco's vision has brought fast growing credit services, along with telephones and other services. They even have different levels of credit cards using their store brand. You can get the regular Tesco credit card that is purple and blue. You can get the Tesco Finest card, which is platinum color. Great identity, especially since the credit cards' color schemes relate to the color schemes of the packages in the store.

For home delivery, Tesco.com is one of the few really profitable dot com businesses in the world for home delivery. The procedure is simple; you go to the Tesco.com website, put in your order and the order gets sent to your local store. The products you order go into carrier bags, into a box, into a van, which is driven to the home, and the box is put on your kitchen floor.

Some Tesco customers joke that they like to go to the store to be seen shopping for the "finest" items, whereas the cheaper items are ordered online so you wouldn't be seen buying them at the store.

Visionary sophistication

High fashion department stores are always looking for new ways to merchandise their products.

Harvey Nichols in the U.K. is a top end fashion store, which combines the best of traditional retailing practices, allowing customers to buy and get expert advice from a variety of specialists in every area.

It recently introduced a food market to sell upscale brands to its well-heeled customers. Its vision was to create its own packaging to reflect the quality image of the store. The packages would not reflect the food itself, but would instead make the shopper feel that the store was giving them something extra, and turn food shopping into a glamorous pastime.

Today, the Harvey Nichols' food packages and labels have a very coherent style. Black and white photos reflect the stores' upscale image, instead of reflecting the particular food product category. They capture the "romance" of the product, rather than the actual product. This appeals to the sophistication of the customer.

The store claims that you won't want to hide the packages in your kitchen cupboard. This visionary, lifestyle packaging, designed by Michael Nash Associates, has influenced other packages in the U.K. market. The packaging helps to build the influence of the store itself, and people go shopping at Harvey Nichols to collect the packages.

So this top end fashion store is certainly style conscious, not only in its clothing, but also in its packaging.

The appeal of the Harvey Nichols package designs is that they reflect the sophistication of the shopper rather than the actual product

The shopping experience

Retail is constantly changing. We spoke about the retail clerk, and how some stores are known for helpful, friendly clerks and you're enticed to go back the next time. In the big discount retail stores, the clerks are paid close to minimum wage. There are still some parts of the U.S. where they can afford to live modestly, and your typical clerk is knowledgeable. However, in many parts of the U.S., the small income puts the clerk in a quick job-changing mode. Go to a store where the typical retail clerk does not familiarize him or herself with the products or the business, and you may not return a second time.

Stores like Tesco in the U.K. and Wegmans and Harris Teeter in the U.S. focus on the shopping experience. Tesco makes sure you have the experience by offering an amazingly wide range of products. Harris Teeter has different destinations around the store relating to different experiences. In Harris Teeter, shopping for health products gives you a different sensory experience from shopping for groceries or shopping

Gourmet departments with many chefs producing quality food for in-store and home consumption make a favorable shopping experience at Wegmans supermarkets

for clothes. Wegmans brings new experiences with an innovative fresh food presentation, many lines of natural and organic products, and innovative brands for fresh and prepared foods. Some refer to this as the anti-Wal-Mart attitude, since that sensory experience includes the background colors in the store, the nature of its walls, and features that make it intuitive where you find your product.

So the big step will be in making the store environment a truly sensory experience with the emotional element to pull in more consumers. The packaging should contribute to the emotional element. We mentioned earlier how Target has done this, and each retailer has this opportunity.

This is the big opportunity for the future, and if you're a retailer, you should take advantage of those sensory elements as a real point of difference. We also know how busy the consumer is today, and the smart retailer will look to move the shopper through as quickly as possible.

Companies like Design Forum use new ideas to create an exciting shopping environment for store chains which is unique and builds on brand loyalty. Combining strategy, research, design, branding, and architecture, it integrates the solutions so that shoppers have good experiences. Prototype stores for chains as diverse as Best Buy and Wild Oats became more accessible and user-friendly with better signage and communication. Wild Oats' natural wood fixtures, cheerful murals, vibrant color palette and premium packaging create a memorable atmosphere.

So what's in store?

There is a future for retailers with vision. Undoubtedly there is a real role to be played by an electronic purchasing system taking place in the home. Press buttons, and the computer at the supermarket comes online. Scan your information, and the product is ordered from your home. Here is where the package is important to carry through the brand loyalty.

The Internet had trouble thinking that the world wants products delivered to their homes. Actually, whether you drive to the supermarket to pick up the product, or whether it is delivered is not that important. The important thing is shortening the consumer's time spent ordering the goods. And engaging the consumer with your brand and package.

Another big change for the future will be in product replenishment.

More and more retailers do not want to pay for products until they sell them. The national brand manufacturer with the deepest pockets is putting out the money and surviving, according to Brian Sharoff, president of the Private Label Manufacturers Association. The other national brand manufacturers who can't afford to put out the money will not be able to stay on the shelf in the future.

To afford these outlays, mergers and acquisitions are moving at a fast pace, and globalization of manufacturing also will offer the affordability. These mergers and acquisitions bring up brand loyalty issues. Which brands remain? The brands that will remain are those with the vision to win the heart and mind of the consumer, with their packages playing a leading role.

Chapter 11
Wal-Mart and everyone else

Start a discussion about retailing, and one name will come up first: Wal-Mart.

When a hundred million people worldwide walk through Wal-Mart stores every week, all you can do is exclaim: Wow, what marketing clout! Or, as Lee Carpenter, CEO of Design Forum, puts it: "When you talk about the future of retailing, it is basically divided into two areas: Wal-Mart and everyone else."

While this may be an oversimplification of today's retail market conditions, Wal-Mart certainly brings out a lot of emotion when you discuss the retail industry. First and foremost, they have the lowest prices or, at least, the lowest price perception. Some Wal-Mart stores may look a bit worn, but customers will endure this in exchange for the lowest prices there. While Wal-Mart may not be noted for its shopping atmosphere, it's the cost of its goods that attracts customers to its stores.

And Wal-Mart has a vision, including extraordinary supply chain economics, strategically placed warehouses and the leverage to get products first, even exclusively for a period of time. No other retailer has clout like that. On the other hand, when Wal-Mart is talking about putting up as many as 800–1200 new stores worldwide, it may be biting off more than it can chew. For one thing, there may not be enough money to go back, retrofit and clean up the older stores. For another, not everybody welcomes the presence of Wal-Mart's size in communities where traditional local stores have had close and long relationships with their local customers and are now fighting for their very survival.

The packaging clout

Wal-Mart also uses its clout in other ways. Products sold at Wal-Mart are packaged the Wal-Mart way. We have heard of one example of this when a manufacturer wanted to put its line of products in boxes with windows. It felt that it would be different from everybody else in the category and convey higher quality. Its design agency tried everything. In the end, Wal-Mart told the manufacturer in no uncertain terms that it wanted the products in blister packs or Wal-Mart would not use them. The manufacturer, not wanting to lose Wal-Mart as a customer, complied.

Our firm had that experience ourselves. One of our clients was required by Wal-Mart to bundle a line of products in larger quantities and in film-wrapped units in order to adhere to Wal-Mart's sales methodology in that product category. Our client agreed, even though the same products continued to be sold to other sales outlets in the customary blister packs.

There is a joke circulating that www does not stand for "World Wide Web" but means "What Wal-Mart Wants"! There could be a grain of truth in this.

The image game

Are you willing to pay extra for package conveniences such as cookies in cups for your car that, on a per ounce basis, cost four times as much as a larger cookie box? If so, you're not a Wal-Mart customer. Wal-Mart's philosophy is based on offering the shopper value, and it has made its success that way. Bob Anderson of Wal-Mart lives and breathes that philosophy. When you speak with him, you can hear him bridle at the discussion of products, the cost of which he considers to be excessive. "In many cases", says Bob Anderson, "it's the marketing person trying to justify his or her position by changing a package and raising its price to more than the product is worth."

Although most shoppers frequent Wal-Mart because of the low cost of goods to be found there, Wal-Mart tries very hard to define the store's point of difference from other mass-merchandisers by communicating an image that goes beyond low pricing – an image of high quality at lower prices. It follows that Wal-Mart packages its private products in ways that convey the impression of being of high quality, equal to advertised brands.

With this goal in mind, Wal-Mart is departing from the concept of putting all store branded products under a single brand identity umbrella, as this might look too uniform and signal cheap products. Instead, in addition to carrying almost every major brand, Wal-Mart competes with these with over twenty proprietary brands in its stores.

These often extensive product lines cover a wide range of categories. They include "Equate" pharmaceuticals, "Spring Valley" vitamins, "Reli-On" health beneficial devices such as glucose meters and "Ol' Roy" dog food. Then there are "Great Value", a low-cost line of food products, and "Sam's Choice" for slightly higher quality food products.

Through these exclusive brands, providing good quality for lower prices than advertised brands, Wal-Mart has succeeded in establishing the perception that these are major brands which are a primary drawing card for loyal and frequent visitors to Wal-Mart stores.

Wal-Mart's private brands use various brand names to position their products as if they were manufacturer's brands but at lower prices

You can rely on it

But Wal-Mart's success with these brands isn't based on luck. The brand and the products are carefully thought out and developed so that equity is built around them within a particular category.

We saw first-hand an example of Wal-Mart's proprietary brand building when we were asked to develop a brand identity and package design system for the exclusive use of the Wal-Mart pharmacy department. The new brand would have to reflect Wal-Mart's commitment to quality and value, and the brand needed to be leveraged eventually into a variety of health solution categories.

By definition, an exclusive brand is generally owned by a store – in this case Wal-Mart – but the brand equity is built through specific expertise within a category. For example, 15 million people in the U.S.

alone require diabetes control. An important category. In fact, 25 percent of all drugstore purchases are attributable to diabetes or diabetes-related complications. So the market is huge.

The initial product line of Wal-Mart's diabetes care products included syringes, lancets, glucose tablets, skin-care products, alcohol preps and, eventually, diagnostic products such as blood pressure monitors, thermometers and other home healthcare products.

After developing the components of brand strategy – vision, mission and values – emphasis was placed on developing a new brand name. Some memorable names were generated, such as Firstcare, Carepath, Wellview, Reli-on and many others. Through consumer research, the selected name was Reli-on.

A brand logo and package design system was conceptualized, refined and implemented to meet the objectives of communicating efficacy and trust.

Collateral materials for Reli-on were focused and mirrored the brand name in the headline: "For diabetes care, I rely on Wal-Mart ... in every way!"

This new and exclusive brand is generating success and, once again, demonstrates how the vision of Wal-Mart brings new highs that go well beyond its reputation for low price points.

Reli-On diabetes care packages reflect Wal-Mart's commitment to quality and value

Efficiency is key

Yet, retailers like Wal-Mart, as well as other hypermarket chains, have a dilemma: How to provide good service at low profit margins on products. How to hire enough people to stock products, answer questions, check out products so that the shoppers can move out of the store without delay, and still offer products at lower prices than other stores.

When the price/value issue becomes a predominant feature, it is clear that something has to give in order to balance the low price points of the products against customer service. With the need for cutting corners to save operating costs that can be passed on for the consumer's benefit, sales assistance becomes scarcer and scarcer even when Wal-Mart TV advertising proclaims the opposite.

The key word is efficiency. To provide room for customer service, today's retailers must apply efficiency in the use of their staff, such as avoiding a lot of time-consuming labor of stacking products on the shelf. Indeed, it becomes the task of the package to replace the sales clerk. More and more, packaging at hypermarkets must provide as much information about the product as possible and it must do this concisely and cogently. More than ever, packaging in the hypermarket *is* the salesperson as there is little hope for the shopper to catch the attention of the disappearing sales clerk to explain the product's benefit.

A case for cases

This, in turn, leads to the use of corrugated shipping containers at many Wal-Mart stores as a means of stacking shelves, instead of stacking individual packages. To make it easier for the store clerks, Wal-Mart developed display ready shipping cases with tear strips that enable clerks to load them directly onto the shelves. These are called PDQ (pretty darn quick) cases. There is no need to remove 12 individual packages from their shipping containers and carefully arrange them on the shelf. All that is needed now is a box cutter to cut the case open, pull the strip, remove the front panel containing information for the supply chain, and just slide the case on the shelf.

When the top section of the front panel is thus removed, the remaining bottom section displays product information for the shopper. This includes brand and product name,

PDQ cases may not please the eye, but they support the efficiency that Wal-Mart seeks on behalf of its bottom line priority

product descriptor, size or weight, the best buy date, and a 14-digit UPC. This may not please the manufacturer who designed the packages to be displayed individually, or package designers with a passion for aesthetics. But life in the hypermarket environment is focused on moving products.

Moving pallets

The cost-effective methods of Wal-Mart's have not gone unnoticed by other big store chains. More and more hyper-outlets, such as Cosco, Home Depot, Best Buy and Lowe's, now use similar display methods that favor product movement at the exclusion of aesthetics.

Shipping cases on industrial steel racks are now the trademark of many hyper-outlets. If this is indicative of future hyper-store layouts, expect it to affect the design of packaging in the long run. Like it or not, while this may make shopping less fun, it will be championed by retail management whose priority is moving pallets of products.

Even the venerable Sears is gradually adopting some of the marketing methods of their mammoth competitors. In addition to their staples of refrigerators, vacuum cleaners and washing machines, some Sears stores now include grocery sections and garden centers and checkout counters near store exits, instead of cashiers at various product-related locations.

The price/value relationship

Aesthetics or not, Wal-Mart and some of its competitors believe that they meet their consumers' needs: Being a one-stop market at geographically convenient locations, having everything most consumers want at prices that are, more often than not, cheaper than advertised brands.

In an economic climate where unemployment is high and stock market earnings low, it is easy to see how cost issues have pushed their way to the foreground of purchase decisions and consumer behavior, affecting everything in the stores, including packaging.

Whatever you may think of hypermarket merchandising methods, what makes their brands especially interesting is that, although the price/value relationship dominates their marketing strategy, some of the hyper-outlets, including Wal-Mart, treat their own packaging with concern about consumer feedback. But their concerns, instead of being primarily aesthetic considerations, are focused on issues that they consider more relevant.

More than a design element

We interviewed Wal-Mart's Bob Anderson and this is what he had to say about some packaging issues:

> Packaging needs to be more than just a design element. You have to go look at functional issues, ethnicity issues, environmental concerns and, most importantly, health information. For example, obesity is such a huge problem in the United States. It's affecting the economy and health and insurance. Packaging could and should play a key role in this. We need to work with nutritionists and, for the benefit of our customers, talk more on our packaging about saturated and unsaturated fat, calories, salt and sugar.
>
> And there are other concerns, practical concerns: Does the package make the product easier, and faster, and cleaner to use? Does the jar drip, when you open it? Is there a safe closure, so when you're driving home and the bleach falls over, it doesn't spill out? Is it safety proof for the children at home? You will see that sort of concern becoming more and more prevalent in the future and playing a larger role in helping the customer make the purchase decisions.

When a senior executive of the most influential leader in the shopping fraternity speaks out about the need for packaging to become more forthcoming about health issues and more concerned about such subjects as packaging functionality and ethnicity, it indicates a topic worthy of further exploration by manufacturers, marketers and designers.

Chapter 12
Packaging as a promotional tool

Marshall McLuhan once said: "Ads are the cave art of the 20th century." This statement may be complimentary or insulting, depending on how you look at it. But products cannot and should not rely only on advertising to influence the consumer. We all know there are many other avenues of promotion, such as public relations, direct marketing and product trial.

The power of the package is often forgotten within this marketing mix, unless the product company has a vision like Gillette. Gillette understands that its packaging, given the importance of its brand, is vital to keeping brands looking their best and communicating the brand essence, which keeps its consumers loyal.

Its brands, such as Gillette, Oral B, and personal care brands such as Right Guard, Soft 'n Dri, and Dry Idea, collaborate and have established a marketing center of excellence for best practice in advertising and packaging.

It understands that packaging has to work harder to break through today's confusing retail market and clearly communicate to the consumer not only the brand essence, but also the specific use of the product and to whom that product is aimed.

In fact, Gillette uses its packaging extensively in parts of the world where it cannot afford to advertise as many brands as it'd like to. Gillette also uses its packaging to play an important role when it can't advertise all the different flavors or extensions of a brand and it needs the brand to speak for itself on shelf.

Some countries in Europe don't enjoy the large advertising budgets that other countries do, or they may prioritize their portfolios differently. As Richard Cantwell, vice-president Auto-ID at Gillette, indicated, his company sells in over 200 countries and territories around the world, and not all of them can support the advertising budgets that one would like to have for all the products. That advertising budget might be set up to promote heavily on blades and deodorants, but not on the shave preparation business. So this is an example of where packaging can become the main promoter.

Absolut, a very visionary package

When you think of a package being central to everything that a brand has stood for, you can see the image of the Absolut vodka bottle. You can't imagine a piece of Absolut communication without some replica of the package.

Why has the bottle become so much a part of the brand? It comes from a vision of several years ago when the entire spirits business had their packages tall, angular and paper labeled.

According to Richard Lewis, vice-president of account services at TBWA Advertising, who has handled the Absolut account for many years, the agency recognized a certain heroics in the bottle shape. His book, *Absolut Book – The Absolut Vodka Advertising Story*, gives great background and photographs of the advertising that it has used through the years.

TBWA recognized the heroics of this bottle, which probably came out of the tradition of Swedish apothecary and pharmacies. And it made the bottle the cornerstone for everything that Absolut does today.

TBWA was faithful and stuck to the idea, starting out with simple, basic ads for the first five years. According to Richard Lewis, this was not one of those overnight successes, but germinated and grew in the early 1980s, until suddenly, by the mid-1980s, it was clearly a very successful brand. The brand was built on simple values, the package being one of them. The agency showed how a bottle need not be boring. It was seeking a wide audience of people to identify not just with the product but also its personality.

According to Richard Lewis, it's easy to put a product on a pedestal, but the difficult part is taking it down. And TBWA was careful not to always put it on a pedestal. It ran an ad during an election year, called "Absolut Primary". Mud was thrown on the package, and the client exclaimed: "How can you possibly put mud on our package?" It was an obvious question for someone in his position to ask. But it was about the personality of the brand being able to laugh at itself and not take itself too seriously.

One would think that showing the same package year after year on hundreds of different ads would be tedious, but the agency looked at it as an organic thing, understanding that the product and the brand, in order to remain interesting, has to grow and take on more personality characteristics.

And that's why advertising and other communications have evolved over the years to take on many different areas – the arts, pop culture, topical areas, music – all sort of become part of the scene, or a scene at which you can laugh.

The bottle is used in the ads in the same way Hirschfield drawings were used in the *New York Times*. The caricaturist Hirschfield put the word "Nina" into his drawings to engage the public into trying to decipher how many Ninas were represented. People look at the Absolut bottle in this way and try to decipher what it represents in these different ads.

ABSOLUT NANTUCKET.

The Absolut bottle shape is always there, although sometimes hard to find. This challenges the viewer and supports the sophisticated brand personality

The most recent ads address the heritage of the product. There's still a bottle in every ad, but the viewer does not see it in the same perspective in which they were accustomed to seeing it. This is intentional and the new advertising program will tell a bit more about the heritage. One ad describes the fact that Absolut is used to rinse out the bottles because nothing else is pure enough.

The product has been used in this way for over 20 years. TBWA has succeeded in doing this by its vision plus the package and, of course, some guts and some endurance.

If we look closely at the bottle itself, the story of how the product is made appears directly on it. It romances the manufacture and reminds people that this product is made in only place, which is Sweden.

In the advertising business, new people come in, want to leave their thumbprint and change campaigns completely. At TBWA, its vision has enabled them to leave its thumbprint by building on what has happened before. A former president of Absolut said: "Consumers are drinking the advertising as much as they are drinking the vodka."

When you think about this type of advertising, it shows how people don't necessarily think of taste by what they're told about taste. They think of taste through what they're told about the brand itself.

When advertising communicates the essence of the brand *through the package*, people start tasting it in their mind. This is a visionary approach.

Like any brand in today's market, Absolut has strong competition, especially from higher priced premium brands like Grey Goose, which are often bought as gifts, an important part of the spirits industry. Fortunately the strong vision of the marketers brought the Absolut package into the hearts and minds of consumers who have had the opportunity to be entertained by the advertising. Hopefully they will continue to do so.

The package as an actor

Advertisers today face a challenge to get their products seen. TV viewers have many channels from which to choose and greater control of the content that they wish to see. Advertising can be zapped and avoided by channel surfers.

But product placement is out there. On many episodes of the Seinfeld TV series, a popular U.S. sitcom about a group of New York friends, strategically placed products and packages circulated in Seinfeld's apartment. These strategically placed items were not coincidental, but relied on the packaging to subliminally remind viewers to shop for that product.

Product placement is an alternative to the traditional method of advertising. It usually relies on the packaging, and is an effective method of developing brand recognition by featuring the product or package on TV programs or in films.

There are many well-known product placements, including Reese's Pieces in the movie *ET* over 20 years ago. The *ET* filmmakers originally approached Mars, the maker of M&Ms, who turned them down. When Reese's received all the worldwide publicity, the Mars folks were livid!

According to *Business Week* magazine, some other well known placements included Junior Mints in *Seinfeld*, Pizza Hut pizza and Nuprin pain relievers in *Wayne's World*, Reebok in *Jerry McGuire*, Ray Ban sunglasses in *Risky Business*, and those "007" films which placed Visa cards, Avis Car Rentals, BMWs, Smirnoff vodka, Heineken beer, Omega watches, Ericsson cell phones and L'Oreal makeup.

So the package is an easy vehicle to place within a film or TV episode because of its potential recognition and convenience of size. Showing the package in films or on TV programs could even imply endorsement of the product and enable marketers to reach their audience, sometimes at a very low cost.

How effective is it? If you use the package in the right place, at the right time, by the right person, the movie or TV viewer can identify with it and react positively to it.

Although not a package itself, James Bond's driving of the BMW roadster generated hundreds of millions of dollars worth of exposure for BMW. The placement helped to drive BMW's business, and waiting lists stretched out for months for the car shown in the film.

It's still unclear how consumers will react to having their favorite shows turn into glorified commercials, so you must be careful with this disguised advertising and its potential to irritate consumers. We don't want to be turned off by strongly branded products in our view during certain dramas.

Watch professional sports on TV and virtual ads show up on the playing fields, although they don't actually appear on the billboard or the field itself.

In the film product placement industry, specialist product placement agencies and ad agencies work together with their corporate clients and studio people to look for the best opportunities for placement. The benefits to the filmmakers, of course, are revenue. But issues that need to be addressed are the artistic integrity of the placements.

Some day, through digital technology, viewers will be able to interact with the product they are viewing and get information about it, and even place orders for its delivery.

Of course, a big advantage of product placement is its far reach, thanks to the ever expanding, global distribution channels for films and TV programs. Think of the international theatrical run that takes place followed by ancillary markets such as home video, pay-per-view, premium cable and broadcast television. These distribution channels are further opportunities for a film to be seen and for a product placement to be observed, and the package is the best vehicle for the job.

Product placement lasts a long time. In the future, products within films will be altered for different regions. For example, a package, which looks one way in the U.S., may have a different look or language in another country, which can effectively be altered digitally.

Many films need packaging as part of their props within scenes, so you should look for the right outlet for placement. The artistic sensibilities of most directors keep them from overcommercializing their films, so hopefully product placement won't become overwhelming.

Hi-tech placement

Today's high-quality image and sound systems offer marketers a way to depict their products with glory. Think of a bottle of Bud from your floor to ceiling, hearing it pour on your surround sound.

Compared to TV advertising, marketers investing in placement are better able to determine the number of consumers they've actually reached.

Some TV producers so badly want to use a particular brand with their scenes that they'll actually pay money to secure the cooperation of the brand in order to avoid copyright pitfalls.

But we better let the package do the job and not expect the actors to communicate product attributes in their dialog. Otherwise there will really be a viewer backlash.

And undoubtedly, placement will be more effective for increasing positive brand attitudes when the viewers have favorable attitudes toward the programming and its characters.

The *New York Times* reported that Sears made a deal with the ABC television network to place products like Craftsman tools, Kenmore appliances and Land's End home furnishings in a six-episode TV series, *Extreme Makeover: Home Edition*. The packaging with brand identification undoubtedly will be prominent, especially in scenes of trucks delivering merchandise. This product placement is estimated to cost Sears over $1 million.

You can see how valuable a product placement can be, since it's not threatened by products like TIVO that enable viewers to zap commercials.

Promoting with special effects

Could a picture of a man on a package shift into a picture of a woman? Or could a hair-coloring product for men show a gray-haired man and then, when the box is tilted, show the hair becoming darkened?

There are many special effects used today, such as light-interfering pigments that shift colors. And some promotions and displays are using lenticular printing and labels for various products. South African Breweries uses a lenticular label on a beer with plastic film and ridges, which shift the image as you move around the bottle. Reebok and General Motors (GMC) also use the technology on labels. It's extra cost, but appropriate for some premium items.

So many things are possible, and you've probably seen packages using holography, which shows shifting images and color.

Perhaps the Internet will drive a major shift in packages, with a new type of package identity revolving around Internet identity. Things may pop up on your computer screen when you select the brand for which you want to shop.

RFID, or wireless network computer chips embedded into the package may also change the way we shop. With "smart" packages, discussed later in this book, self-checkout is made easier at the super-market. But don't get carried away with the potential of the chips – including sound. Your product can be promoted – subtly we hope – as the shopper walks by the shelf.

The follow-through

But with any promotion for any product, the communication elements must work together. Marc Rosen tells us the story of a fragrance designed for Halston some years ago called Catalyst. The Catalyst bottle itself was an abstraction of women with Halston gowns, which were consistent with the image. However, the advertising and the promotion for some reason used a model on the prairie showing a split fence and cowboys, the antithesis of the whole idea of the Halston brand.

The promotional follow-up was not synchronized. The fragrance failed.

Customized packaging

Some dictionaries define packaging as a number of items bundled together as a unit. That's a good definition when you think of how mass-produced packages have enabled products to be marketed, stored, shipped and displayed all over the world.

But how about a vision where the package becomes targeted toward a particular consumer?

With today's technology, it's possible to target even further on a package by integrating a particular store message with the brand on the package. For example, a product from Kraft could have a message from your Kroger supermarket.

Or you could target toward a specific part of the country, or toward an ethnic market.

Perhaps the back panel can have specific messages for each of these regions, or specific messages for a particular audience.

This is a big opportunity in the future for you to be in the forefront, with packaging that addresses the individual. There will be printing and distribution challenges, but once these are overcome, the vision will be inaugurated. For example, localized packages with localized languages could be used, targeting the parts of the country that would affect those particular segments. This is really meeting the consumer more intimately.

But it takes vision, and it takes a company willing to put the systems in place to support innovative ideas. It's more than just having great ideas, it means having the vision to get them implemented.

One day we might see packages that change within the neighborhood stores. Today it's probably easier for a small company to make decisions that do revolutionary things like that, but it's a big opportunity for any company.

Mass customization

Let's look at how some companies are delivering features to consumers that are more customized.

FedEx has a packaging lab and will work with companies to develop packaging for whatever product they have. For Harry Potter books, a custom package was designed for book insertion, handling, delivery and co-branding.

This custom designing is done for companies to benefit not only the company but also the package suppliers, in this case, FedEx's ability to deliver the packages. With the Harry Potter books, FedEx had partnered with their publisher and Amazon.com to home deliver a quarter of a million books the first day *Goblet of Fire* was released. These preordered copies were delivered on time. Amazon and FedEx co-branded and, with their vision, also see future directions in co-branding programs based on strategic alliances.

FedEx can expand its packaging lab to come up with new solutions and have a personal relationship with customers to help them with their needs for their own packaging.

Gillette is another example of a company that wants to communicate with its consumers and their specific usage habits. Its strong interest in RFID chips may eventually head them into a future of intelligent personal products. If it's toothbrushes, and there's a way Gillette can make the brush head more personal to the consumer by telling them when it needs to be replaced, that they are brushing too hard with it or that they should use a different consistency because they're a very aggressive brusher, Gillette wants to impart intelligence into its product to be played back to the consumer. This would make its product more customized, more personal and its brand relationship more intimate.

Use your senses

Should you show your product? This is a big decision when designing a package.

With today's technology, printed photographs or illustrations on packages can look superb. And of course, for products like cake mixes, where only the powder is in the box, the decision is easy – show a great picture.

But on many products, such as fashion accessories, a picture just won't replace looking into the box and seeing the actual thing. Repeatedly, research studies have shown that an attractive product will always appeal more than a picture of the product, or even appeal more than a fancy gift box that doesn't reflect what's in the box.

The problem is that it's usually more expensive to have an open carton or a carton with a window. But if your product really looks good, don't think twice. It'll be worth it in the long run. People want to see what they buy.

There are instances where box windows are used in the wrong manner. We remember a Fisher-Price brand toy set filled with little trains and houses that were all sitting in the box. The window showed all these parts, which looked jumbled. A year later, Fisher-Price was smart enough to close up the box and show a brilliant photograph of all the pieces assembled and looking like a fun kids' town with little houses and trains in line. The closed box was far superior to the open box in this case and undoubtedly more successful at retail.

The package is the product

There's an old expression about the package being the product. This is especially true with many food, drug and cosmetic packages where the product itself might just be a powder, liquid or unglamorous item. The package makes the sale.

An example of a good use of bottles is Fleischmann's vegetable oil spread. A pearlescent olive green bottle is used which immediately conjures up thoughts of olive oil. It also makes a great point of difference for their vegetable oil spreads. On the shelf with other liquid margarines packed in yellow bottles, this olive green bottle reinforces the ingredient and communicates the brand persona of the product, making it a winner for both the retailer and the consumer.

What can you do with a wholesale package of fish? Even products for restaurants can get in the act. A company called Honolulu Fish Company worked together with Weyerhauser to ship exotic Hawaiian fish to fine restaurants and hotel chains in the U.S. To have a big impact, they put a holographic cover made by Proma Technologies over their entire corrugated box. The effect was substantial and Weyerhauser did tests which discovered that the reflectivity of the holography actually kept the product inside cooler.

So the restaurants all remember the Honolulu Fish Company because it brought a sensual appeal to its shipping box and it wasn't only cosmetic. In fact, restaurant employees took the empty boxes home to use as coolers and sunshades in their cars.

The visionary brewery

Many factors affect the way packages are used in different parts of the world. Argentina's closed economy before the 1990s enabled their local products and packages to be leaders in their markets. When Argentina opened its market, many brands were forced to compete against one another, especially since external sources became prevalent at retail.

Quilmes, the largest beer producer in Argentina and in the combined southern cone markets of Argentina, Bolivia, Chile, Paraguay and Uruguay, had to compete against many other beers from different parts of the world.

The year 2000 was approaching, and Quilmes saw an opportunity to exploit new label technology. Together with the designers from Interbrand, Quilmes recommended using the beer packaging to promote and reflect Argentina's colorful culture.

Starting with packages in the new millennium, the Quilmes beer bottles incorporated a new design based upon the paintings and posters that the company had used in its campaigns in the past. These posters were well known to Argentine consumers and had become cultural icons, so the recommendation was to use the poster artwork as full plastic wraps around the beer bottles.

To make the vision complete, it was agreed that the bottle graphics would be used for only one year, and would be changed to new designs after that year. It was made clear to the consumers that these designs were limited, which made the bottles more exciting to purchase. With the new poster graphics reflecting the history of the company, the vision was off to a good start. The idea was so popular that the poster graphics were also adapted to a 5-liter beer can.

The following year, the images on the bottles reflected Argentine culture through images of tango dancing, car racing, the 1960s and World Cup football. These are the popular passions of Argentina, and signify the way that Quilmes beer reflects the Argentine populace.

The next year, the beer bottle graphics reflected Argentine craftsmanship, using photography to show leather, metal and wood – materials that come from nature and the land.

The following year, well-known celebrity personalities, or national idols, were used to reflect the national culture.

Now the beer reflects Argentina and its culture from rich to poor, young to old. The bottles can be found in supermarkets, liquor stores and even flea markets, since they have become collectible items within the country.

And the strategy, which started out for the new millennium, is so successful it has been extended to future years.

Quilmes beer bottles, reflecting the Argentine culture, have increased sales and have become popular collectables. The designs change annually

The multi-sensory experience

Some companies today are looking for ways that their brands can deliver more multi-sensory experiences to consumers at the point of sale. This means the ability to see, feel, touch and perhaps try certain products out at the point of sale. There's no doubt that consumers like to try out and experience products before they buy them and this can only help the producer sell the product.

Let's take shampoos. In the future, consumers will like to sample skin care products, or perhaps smell new shampoos rather than opening up the bottle.

So there is an opportunity to address consumers' desires to experience the product or get that multi-sensory experience a little bit more to the point of sale than it is there today. The future will be more driven by consumers' needs and delighting consumers and surprising them, and the brands that are able to do that will be the winners.

To be successful, the retail environments should deliver more entertainment value through these brands delivering these experiences. What a great opportunity for the future. We know how stores like Starbucks have been able to create a consumer experience by setting up their outlets in a way that consumers use and buy the product within that coffee experience.

And a trend will be for products to do more than just deliver the benefit that is expected. In the future, the successful shampoo will deliver more than just clean hair, but also the experiential needs, or the experience of the product. For example, that might mean more emphasis on scent and the way the product actually feels in the shower, and even the attitude conveyed by the design of the package and the advertising. This adds up to more emotional benefits for products and for brands in the future.

Chapter 13
The package on the road

What is the first envelope on your desk that you open when you arrive in the morning?

There's a good chance that it's the FedEx white envelope that arrived before you did. Because that FedEx Express white corporate packaging is considered by FedEx to be one of their brand power applications, according to Gayle Christensen, director of marketing and global brand management for FedEx.

The FedEx brand is made up of several companies, including FedEx Express, with purple and orange logo colors, FedEx Ground, purple and green, FedEx Freight, purple and red, and the Custom Critical purple and blue colors.

But we all know the white package, or the white envelope, which stands out from all those other packages on our desk. It represents the FedEx brand by its convenient opener/closure, and the fact that it's seen on thousands of desks and warehouses around the globe. You can see it through the clutter on your desk, and it's consistent with other power applications in communicating FedEx's brand attributes.

The white corporate packaging is an important part of the company. The packages are convenient because of their shapes, the many different sizes of envelopes, including architectural drawing tubes. Some consider it fun to pull off the strip that opens the packages.

But the most important attribute is that the sturdy package, its white color and strong logo gives them priority. You see it first. This is the power of the brand speaking. And there is a consistent look in the design of the packaging that matches the vehicles that you see in the street, the signage you see in the world service centers and the aircraft in the sky.

And all of them work together to communicate that the company has its act together, and communicate that you can anticipate the value and the kind of service you're going to get from FedEx. This is its vision.

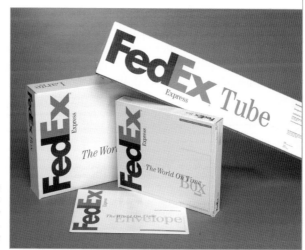

FedEx packages are the easiest to spot on your desk and they coordinate with its branding on the road

Have you noticed how home delivery has changed? In some small towns, the local furniture store uses its own truck to bring your purchase to your home. But today when you see a delivery truck with a well-known product brand or store brand logo, it's usually bringing that product, brand or store brand from the distributor to the retailer's location, not to your home.

The shopper has many choices. She can go to the store, the catalog, the Internet or the telephone. She can look over the merchandise and pick it up or have it delivered. The retail store saves costs and expedites deliveries through delivery services such as FedEx, UPS and others.

FedEx, for example, has been visionary in responding to the delivery market. As part of its ground network, it has a company called FedEx Home Delivery, which was created for residential deliveries. Undoubtedly it was also created because of the increasing need for Internet delivery of packages. Its deliveries cover many countries and 100% of the U.S. and its efficiency is such that customers can track their information on their website.

From point A to point B

Ask marketing people what they remember from the movie *Castaway* with Tom Hanks, and they'll often mention all the publicity that FedEx received in the movie. Tom Hanks, a FedEx guy, was on a FedEx plane when it crashed, and FedEx trucks and packages were used throughout the movie.

In fact, when you think of either FedEx or UPS, you probably think of a guy in shorts arriving with a package delivery, and often the package itself is memorable.

Movie people come to FedEx wishing to place their products or vehicles in their movies or TV shows. They come because of the FedEx brand and what it says.

Packaging is crucial for the courier services. You can imagine some of the packages that FedEx moved from point A to point B when they made what they referred to as "unusual shipments":

- The sets, props and sound equipment for Michael Jackson's 1992 "Dangerous" tour were flown via FedEx from Los Angeles to Frankfurt.
- On separate flights, FedEx has shipped Armand Hammer's art collection and bottled dirt from Houston.
- When a storm struck New England, a fast-food giant chartered a FedEx plane for an emergency supply of special sauce.
- FedEx has shipped pieces from Catherine the Great's collection in Leningrad to Memphis.
- Japanese customers have hired FedEx to ship them planeloads of fresh cherries.
- FedEx has shipped two ancient sarcophagi (stone casings for mummies or coffins).
- A North Dakota bait wholesaler ships worms, minnows and leaches to bait shops across America via FedEx.
- FedEx delivered 12 decorated eggs for a White House Easter egg roll.

The package engineers must have had a blast with some of those products.

As a further delivery service, FedEx and its own packaging lab works with companies to develop packaging for whatever product they have. And as visionaries, FedEx sees a variety of directions for the future. Some future directions include:

- co-branding programs based on strategic alliances
- expanding the packaging lab to come up with new solutions
- working in a more personal relationship with customers to help them with their needs for their own packaging
- forming specific relationships with customers for customized packaging
- helping customers with their packaging development needs.

The big payoff for delivery services will be:

- developing the proper sized package configurations
- using technology through efficient tool systems for the people who manage the deliveries
- organizing the vehicles to handle efficient packaging.

The all-around business

How is it possible to compete effectively with a delivery service like FedEx that can handle all those unusual type of shipments all over the world? Well, there are many competitors, and FedEx goes head-to-head quite often with UPS. UPS is actually larger than FedEx, but the public is so used to seeing the brightly colored FedEx packages around that they probably don't realize it.

At one time UPS just shipped things. Its vision is much wider now. From its advertising campaigns, you can see that it now handles ware-housing, e-commerce, customer services and your logistics if you are a large company, and, by the way, they do a *lot* of shipping.

UPS prides itself on its functionality in terms of package logistics. Its "smart" coded labels simplify the loaders' jobs so that what used to be a three-hour job takes twenty minutes in loading the truck. The packages are automatically sorted, and the label tells a preloader where to put the package in the truck. Sorting and loading activities revolve around this label itself. This is pretty important since they handle over 13 million packages daily around the globe.

The UPS package lab in Addison, Illinois develops and tests packaging for every item imaginable, performing drop tests, compression tests and tests to show when items may sustain damage. The lab places a vibration tester in the UPS truck or plane, and then returns it to the lab to play back bumps and vibrations so the engineers can study the effect on newly designed packaging.

These days e-commerce has brought us smaller packages. The Internet enables people to order and have delivered every product and size imaginable.

According to the magazine, *eCompany Now*, UPS has an e-commerce sales force dedicated to persuading business customers to integrate UPS's package shipping data into their websites. A great competitive edge for both UPS and the business. You simply click on your store's website and you're instantly connected to UPS's database to get the status of your delivery. This saves the step of getting a UPS tracking number from the store and then heading to UPS.com for the information. More direct, more time saved, and a happier consumer.

In the article, Merrill Lynch analyst Jeff Kaufman says: "The universe of people who can provide both information visibility and transportation across the world is a very small fraternity." You can count them on one hand: UPS, FedEx, Deutsche Post (which owns part of DHL), and perhaps one or two others.

Probably one of the biggest things separating UPS from FedEx is the warehousing. "FedEx sticks with its strengths of moving things quickly and coordinating those moves, not having a lot of parts sitting somewhere," according to Tom Schmidt, head of FedEx's worldwide e-solutions.

But UPS's network of warehouses enables them to provide other services for customers, including technical repairs and goods assembly.

It's important for courier services to visualize their daily processes for efficiency. Years ago UPS started conducting time and motion studies on delivering packages. These get updated every few years. The drivers now follow over 300 scripted movements to make sure there are no wasted steps. For instance, when they enter the truck they insert the key with their right hand while pulling down the seatbelt with their left, then release the hand brake and the clutch at the same time.

Sorting it all out

How many people does it take to sort millions of packages and containers? Actually, not very many, according to *eCompany Now*, if they use smart labels like those used on UPS packages.

For example, UPS has a 550-acre distribution hub in Louisville that is amazing. It's filled with conveyor belts, slides, corkscrew chutes, digital cameras, bar code scanners and tilting trays. Containers are rolled in and unloaded and sorted for their destination automatically.

"The whole premise is that the package comes in smart so it can sort itself," relates vice-chairman Mike Askew in the article. Each package is weighed, then scanned by a digital camera reading its bar code label which contains all the information about the package's origin, destination and the speed it needs to get where it's going. The computer determines the optimum path to take through the sorting facility and instructs a row of black rubber pucks to slide across the belts and push the box down the appropriate chute.

Eventually the box makes its way to a final slide assigned to one of thousands of destinations. And this is done accurately without people.

section four

trends in
visionary
packaging

Chapter 14
Brand
packaging

How long will it take for marketers to look to packaging in the same way they look at manufacturing and production costs?

The marketing environment is changing constantly. How will your packaging function in this changing environment? Will your packaging stay unremarkable and safe in the next five or ten years from today? Or will it take its cue from the electronic industry and become the dynamic champion of progressive marketing that it could and should be?

Will marketers consider package development costs as an expense, rather than as an investment in the future of the brand?

Four issues affecting package design will be central issues in the next decade:

1. The emphasis on branding
2. The role of package design agencies
3. The proliferation of mass marketing
4. The significance of e-commerce.

The emphasis on branding

There is no question that no single issue in recent years has initiated a more profound shift in the marketing of consumer goods as has the emerging emphasis on *branding*.

Branding, as we explained before, is the means of establishing in the consumer's mind a distinct personality by which the consumer will recognize and remember a product. Packaging can communicate this personality through a variety of visual cues.

There are numerous examples of cues that are recognized by consumers all over the world; having been properly and carefully maintained over the years, they provide an incalculable value to their owners.

Well-maintained brand logos provide consistent brand recognition and communicate commitment to consistent product quality

Examples of such cues represent a variety of opportunities, such as:

- *a trademark* – Lacoste, Xerox, Lipton
- *initials* – BMW, BP, M&M
- *an icon* – McDonald's, Apple, Dubonnet
- *a signature* – Kellogg's, Cadbury's, Johnson & Johnson
- *a symbol* – Shell, Nike, Quaker Oats
- *a name* – Danone, Gillette, Heinz
- *a personality* – Uncle Ben's, Betty Crocker, Morton Salt girl
- *colors* – Kodak, Campbell's, Bayer Aspirin
- *a package shape* – Coca-Cola, Chanel, Odol Mouthwash
- *an overall design program* – Lego, FedEx, IBM.

Or it could even be a combination of several cues, as in the case of Coca-Cola's combination of bottle shape and signature.

Brand identification identifies your products wherever they are sold, be it on packages that are neatly arranged on supermarket shelves, or not so neatly on steel shelves in hypermarkets. They may appear on your TV screen or your computer monitor or flash by on trucks rolling along the highway.

"Brand identity has to feel like a friend", says Howard Schultz, chairman and CEO of Starbucks Coffee Company in the book *The Future of Branding*, edited by Rita Clifton and Esther Maughan. "People have more choices today than they've ever had before and so a brand must be a bridge of trust to the consumer."

A cruel beginning

Branding is not the recent phenomenon that it is sometimes perceived to be. Actually, it has been around for several centuries. The word is a derivative from the Norse word *brandr* which means to burn and survives as the custom among cattle farmers to define ownership of their livestock by burning identifiable marks into the animal's flesh.

This early and seemingly brutish custom has benefited by evolving, over time, into the more sophisticated application of the word "branding" that we use today. Most of us don't think any longer about cattle when we refer to "branding" as a method of identifying merchandise. But it maintains its connotation as a means of communicating ownership and recognition and defining trade dress.

A key marketing tool

But branding as an essential marketing tool has come into full bloom only relatively recently. When the plethora of corporate mergers emphasized the critical need for sorting out, identifying and differentiating products and services and presenting them to the consumer in a more organized manner, branding took on a life of its own.

The need for help to manage an initially unknown component of marketing soon produced a flourishing market of consulting agencies all over the world, specializing in the development of branding products and services.

By now, branding has become a marketing hotspot that radiates into every niche of the communication media. Even ad agencies have begun referring to advertising campaigns for retail products as branding campaigns. Seminars on the subject of branding debate the pros and cons, objectives and procedures of brand identity programs. An avalanche of book and articles about branding have invaded our professional libraries in recent years. Branding has become the glue that binds corporate entities and their various products and services together.

In the book *Brands – The New Wealth Creator*, Tom Blackwell points out that:

> brands … are now recognized widely as business assets of genuine economic value and as such have attracted the attention of a much wider audience. Brands are now center stage: they drive major mergers and acquisitions; they appear frequently in the balance sheets of their owners … they

have changed irrevocably the way in which many major companies organize and run their business.

OK, but what has this got to do with packaging?

Branding the package

When the importance of brand identity takes on such a center stage position in the marketing of consumer products, it is bound to spill over to the brand's packaging. Brand and product managers soon began to recognize the need for visually identifying and clearly differentiating one product from another.

It's crucial to understand that the visual identity of packages plays a critical role far beyond the store shelf. The package follows the consumer into the kitchen, the bathroom cabinet, the breakfast table, the workbench, the garage or wherever the product is used. In fact, this follow-up is more critical for maintaining a favorable and lasting perception of the brand than when you pick the package up in the store.

Brand recognition on the store shelf follows you to your home

Add to that the growing influence of Internet shopping and you begin to realize why the emphasis of brand packaging should be treated as the centerpiece of any marketing program for retail products.

Enter the package designer

At the beginning of the 20th century, when products began to be sold more and more in packages instead of being picked out of barrels or handed to the customer from behind the counter, shoppers began asking their grocers and druggists for products by brand names and retail merchants, realizing that packaging was not merely a container holding their products, began to recognize it as a powerful communication instrument.

It did not take long for a few designers, intrigued by this new method of product distribution, to realize that offering their services to make brands more memorable and packages more informative and attractive would gain them the reputation of being the creative harbingers of this newfound sales medium.

Starting in the 1930s, package design pioneers, such as Raymond Loewy, Jim Nash, Frank Gianninoto, Bob Neubauer and Karl Fink, became household names in the rapidly flourishing packaging industry.

Packaging = Branding

Then package design pioneer, Walter Landor, took package design to the next level. He prophetically equated package design with *branding* of products and product lines. Packaging *is* branding, he pointed out. Packaging identifies the brand. It spotlights the brand. It creates an association with the brand's products. It promotes confidence in the brand's products. In the consumer's mind, the package *is* the product.

It changed the attitude towards packaging forever. Great thinking! Great progress!

But not everyone was listening

A few marketers and designers paid heed to this philosophy, but not many. In the heady postwar days of rapidly proliferating supermarkets and self-service outlets, shelf impact was everything. All you needed was a tachistoscope, an instrument that exposed products in brief flashes to a group of respondents to verify brand recognition and findability. This assured marketers that their packaging graphics were bigger and more blatantly aggressive than those of their competitors. Convinced that they had achieved the ultimate package design, product managers patted their backs in self-satisfaction.

Outscreaming the competitor at the point of sale was the name of the game. It dominated all other criteria. Positioning the package as an integral component of an overall branding architecture was seen as pie in the sky and of little interest in those days.

How ancient all this sounds today. It took the frenzy of corporate mergers during the 1980s and 90s to shift strategic emphasis from day-to-day product sales to looking for ways of building long-range brand strategies. Positioning brands and brand lines inherited in these mergers now became key criteria.

With this shift, packaging began to start playing a new role. Marketers began to realize that packaging must be an integral component of the marketing strategy for retail products.

Undervalued and unappreciated

As packaging began taking center stage in retail strategy, a market in package design blossomed with it. Meanwhile, educational institutions had not yet caught up with the potential of teaching package design. The professional expertise of package design requires a wide perimeter of knowledge, information and hands-on abilities. No such curriculum existed at that point of time.

The professional skills expected of the brand identity and package designer go well beyond talent in graphic and three-dimensional design. Their professional expertise requires combining their creative talents with in-depth familiarity with numerous business and intellectual aspects. These include an understanding of marketing, marketing strategy and positioning, sales psychology, consumer research methodologies and much more.

Then there is the expectation that the package designer is thoroughly knowledgeable and experienced in a wide range of technical issues, such as pre-press procedures and various printing techniques, familiarity with packaging materials and packaging machinery… and the list goes on.

When you add to this the designer's need for intimate familiarity with computers and a large number of computer programs, you will understand that the profession of package design demands a full plate of business and technical know-how that has few equals.

Why then would anyone fail to take full advantage of the package designer's professional skills and capabilities in helping to create one of the most critical components in marketing retail products?

Package designers complain of being the most undervalued and unappreciated participants in the marketing process. If branding and packaging are to continue to be a key factor in future marketing – and we have no doubt that they will do so – it seems obvious that the relationship between the marketer and the package design practitioners is one of several critical problems that needs to be fixed.

Chapter 15
The obvious is not always obvious

Certainly, we can all agree that marketing a product requires packaging that will excite the consumer. This therefore should be an obvious and acknowledged objective for smart marketers.

This assumes giving the package development cycle sufficient time, sufficient attention and, yes, sufficient budget. Unfortunately, some brand and product managers, especially when fresh out of college, don't always understand this. They don't realize that the creation of exciting packaging requires a design strategy that must be no less meticulously focused than the marketing strategy itself.

But don't blame the brand and product managers for this. Blame the professors who teach marketing at our colleges and universities.

In most college-level marketing courses, packaging is not much of an issue. Just look at some of the marketing literature. If you are lucky, you may find two or three pages devoted to the subject of packaging. Most of the professors who write these books have no experience or interest in packaging. They don't seem to understand how critically important packaging is to the marketing process of retail products.

For most of them, packaging is a minor detail of teaching marketing strategy, deserving only minimal time. They seem to view packaging as a design discipline, taught by the likes of the Pratt Institute in New York or the Art Center in California. They don't appear to realize that branding and packaging are an integral part of managing marketing in the retail trade.

Some professors will acknowledge that while packaging should, indeed, be part of their marketing curriculum, there is so much teaching territory to cover that it leaves little time to spend on this subject.

Packaging is thus sidetracked in the marketing curriculum of most higher education institutions. If we acknowledge that "the package is the product", this should be problem number one that requires fixing.

Fortunately, some of this shortsightedness is gradually disappearing. A few colleges and universities are now including full-scale packaging courses as part of their marketing programs, devoted entirely to managing branding and packaging. Among these are those led by Professor Allen Glass at the Kellogg Graduate Business School of Northwestern University in Chicago and Professor Hope Corrigan at Loyola College in Maryland. In a similar vein, the Pratt Institute's New York City facility offers postgraduate package design courses.

In addition, some companies, such as Kraft and Procter & Gamble, have brand and package design-savvy people as part of their marketing teams.

But as late as the 1990s, it was a narrow-minded and damaging assessment that sent marketing students out into the business world with a narrow perception of packaging that limited their marketing productivity. More often than not, package design decisions were made by "gut feel", like buying a tie. It was simply: "I like it" or "I don't like it". Forget package design criteria, if clearly defined package design objectives even existed. Decisions were made by instinct and narrow personal tastes.

Only later do the hard knocks of business experience bring to light that packaging is, indeed, a key marketing element in many categories and that achieving an effective package design requires every bit as much attention, energy and budget as any other segment of the brand's positioning strategy.

A stalemate in communication

This brings us to problem number two that needs to be fixed.

Despite the belated interest in package design as an integral part of marketing strategy, some disconnection between a package designer's creative expertise and some marketers' casual style of treating package design decisions still exists. When it does, it's a missed opportunity, the blame for which must be equally shared by marketers and designers.

The marketing challenges of the next decade will be so decisive for product growth – or even product survival – that *every* element of the marketing spectrum will demand equally serious understanding and attention from marketers and the service organizations they use.

But for this to happen in the future, to acknowledge the importance of package design as one of the key components of retail marketing, two initiatives have to be set in motion by both marketers and their design agencies:

1. Designers must do a better job of identifying the breadth of their ability to contribute to their client's business. They must become more proactive by taking the initiative, with ideas and suggestions, instead of waiting for the client or their ad agency to contact them.
2. Marketers and their ad agencies must be willing to include the design agency in their decision-making process, starting at the very beginning and continuing throughout the strategy development process – not after all the strategic decisions have been made.

How can this be achieved?

The designer's side of the issue

To achieve a more integrated relationship with their clients, designers and design agencies must become more proactive. This needs a change of tactics.

Not noted for their proficiency in self-promotion, most designers publicize their activities by occasional mailings or "leave behind" brochures showing their work. A valiant effort but, more often than not, an ego trip that falls short of effectively communicating their true ability to contribute to their client's business objectives.

More recently, most designers and design agencies have launched websites that serve the double purpose of promoting their experience in brand and package design by displaying their work to other clients, as well as using the design of the website itself to communicate their design skills.

Good try! But not well targeted!

Considering today's superabundance of websites on the Internet, it is uncertain how often a brand manager who is *seriously* searching for a brand identity and package design professional would rely only the Internet to find one. Similar to TV commercials, websites are a buckshot method of promotion to achieve "hits", but the potential of a serious business connection growing out of this is minimal.

Even if this method of communication results in an occasional contact, for a potential client to *really* identify the designer's capabilities and comprehend them, there must be personal communication.

No organizational support

Unfortunately, package designers and design agencies in the U.S. get little help from professional organizations that should be plugging the package designers' marketing expertise as a critically important component in strategic development and helping clients to make meaningful connections with designers and design agencies.

There isn't even an official listing of designers and design agencies for anyone seeking such services, except a small listing that appears in the end-of-the-year issue of *Brand Packaging*, a professional monthly publication.

The once proactive organization representing professional package designers in the U.S., the Package Design Council International (PDC), faded out of sight some time ago. Fierce competitiveness among its members and lack of organizational leadership blunted the council's effectiveness.

What was left of the organization changed its name to the Brand Design Association and merged with the American Institute of Graphic Arts (AIGA), an organization of design professionals that is very effective in promoting design excellence in general, but has shown little interest in and understanding of the profession of package design.

The Pan European Brand Design Association holds frequent conferences, providing its members with opportunities to discuss pertinent business issues

In Europe, the Pan European Brand Design Association (PDA) has taken up some of the slack, Twice-a-year conferences address various packing issues and promote the skills of its members, located in 18 European countries and a few countries outside Europe. PDA's website and an annually updated brochure identify PDA members for anyone in need of brand and package design services.

The new name

As the world and its marketing methodologies have changed during the past two decades, so has our vocabulary. Stewardesses have become "flight attendants", secretaries are "administrative assistants"; used car lots now sell "pre-owned" cars; using contraceptives is "planned parenthood"; ad agency account managers are "marketing consultants" and stockbrokers "financial advisors".

Designers are no exception. With branding having become *the* engine that drives today's marketing strategies, package designers, anxious to gain their client's respect for their marketing savvy and eager to cash in on their clients' preoccupation with branding, were eager for a new identity. Being labeled a "package designer" was like being a descendant of the Neanderthal generation. Upstanding design consultants no longer wanted to be identified by such an outmoded title.

Exit the package designer – enter the brand design consultant.

What's in a name?

Unfortunately, this change of identity from package designer to brand designer has had the unplanned consequence of demoting package design to a second-class status vis-à-vis branding – a commodity. A required component of marketing products, but less significant than branding and, with plenty of practitioners to choose from, easily available at lower cost.

So what has been achieved with the name change? Shakespeare put it this way:

> What's in a name? That which we call a rose
> By any other name would smell as sweet.

Let's face it. Have package designers, since their change in identity, really gained greater respect from marketers? Are they called to the strategy table more often now that they are *brand design consultants*?

Unfortunately, the answer is in the negative, Package designers are not included at the strategy table any more than before their identity conversion. There is an obvious disconnection.

This is problem number two. For the benefit of both marketers and designers, this disconnection is badly in need of fixing.

Chapter 16
The designer as a business partner

Some time ago, the Pan European Brand Design Association sponsored a survey titled "What Do Clients Want from Brand and Package Design Agencies?" As the survey's title implies, the members of that organization wanted to find out what clients expected of them.

The results were revelatory.

While praising the designer's creative skills, the client's perception of designers was that they were:

- Technicians rather than visionaries.
- Interpreters rather than creative innovators.
- Artistically creative but lacking an understanding of their clients' strategic needs.
- Developing design directions based on the marketing positions handed to them by their client.
- Loath to challenge their clients in the way ad agencies often do.

Ah, there's the rub – *challenging the client the way ad agencies do!*

Being a business partner

It is true that advertising agencies have achieved a position that design agencies, for the most part, have not.

As a marketer you often lean heavily on your ad agency to develop strategic criteria for your brands. Thus, your advertising agency is not merely a creator of ads and commercials and a media buyer. Your ad agency is, in fact, part of the company *hierarchy*. A visionary associate, your eyes and ears, a shaper of your brand personalities. Your ad agency is, in fact, your business *partner*.

Design consultants, on the other hand, are most often perceived as project-by-project design suppliers, or worse as "vendors", hired to adapt package design to the marketing strategy that you and your agency have determined.

Needed: a more holistic approach

If, in the wake of the emphasis on branding, package design has become a dinosaur, perceived as a commodity, fulfilling a need but undeserving of more than minimal budgets, it behoves both you and your ad agency to reexamine these perceptions.

That's problem number three. If you accept the premise that the package is a critical component of your marketing strategy, that the package is vital to the brand, it does not make much sense to treat the package designer like a handyman, to be called upon only when you need to fix a problem with your packaging.

In the decade ahead, when every marketer is likely to face increased competition from mega-outlets and the Internet, the need for developing criteria focused on energizing consumer interest in your brands and products will require a more holistic approach. It will require a mind-set that will take advantage of exploiting *every* source of creative input – whether it concerns your advertising, your promotions, your website, your packages or any other element of the brand strategy.

If this seems like a momentous change from your previous decision-making procedure, keep in mind that there are momentous changes taking place in retail marketing. These are not likely to go away. Marketing methods, lifestyles, the sales environment and competitive pressure are bound to keep changing year after year.

With that in mind, keeping your design agency at arm's length until an intimate entourage of internal staff and ad agency personnel have finalized all strategic issues makes no sense whatsoever.

Failing to include your brand and package design consultant when making vitally important strategic decisions narrows your perspective on being unique in your category and deprives you of the opportunity of benefiting from the designer's experience and visionary capability that could lead to unexpected solutions.

Consider just one example:

An almost missed opportunity

A few years ago, Breyers Ice Cream, then a regional brand of ice cream, available in a few states along the east coast of the U.S., approached us to help it relaunch its line of ice cream flavors with the objective of achieving *national* distribution. To achieve this, redesigning the Breyers packages was a key requirement.

The direction we recommended featured a bold new Breyers logo and large, close-up photographs of mouth-watering scoops of ice cream against a simple black background.

Black ice cream packages? Breyers' president and his marketing team were shocked. No other ice cream producer had ever done anything like this.

It is true that, at that time, virtually all ice cream manufacturers – yes, even Breyers – marketed their ice cream in white containers with lackluster, stock ice cream photographs, provided by the package converters at little or no cost.

The old Breyers Ice Cream packages adhered to the traditional use of white packaging and stock photography for dairy products

But we were convinced that following this category tradition would lead nowhere. To stand out in a market saturated with boring, look-alike package designs, we were convinced, would not gain Breyers its stated goal of gaining national distribution. What Breyers needed, we advised them, was to show creative leadership by means of packages that had a point of view.

The new designs represented a revolutionary change in the dairy category from a visual and investment point of view. The graphics were different from all others in the category and the ice cream illustrations, using skilled food photographers, would not be cheap.

But the packages would communicate higher quality than any ice cream marketed at that time.

The visionary new Breyers Ice Cream containers broke with tradition and provided Breyers with the platform for national distribution

The black background, we argued, calls attention to Breyers' bold new brand identity and highlights the beauty of the mouth-watering scoops of ice cream. It's a way of making a powerful statement in an otherwise lackluster market. After all, what are ice cream shoppers looking for? Mouth-watering ice cream, what else?

We did not minimize the fact that it was a major decision on the part of Breyers' management. But we stood our ground. We were convinced that, to gain national distribution, this was the only way to go.

Persistence pays off

It took several rounds of extensive consumer research that clearly favored the new designs to convince Breyers' management to launch a limited trial in a regional area before daring national rollout.

The trial run of the redesigned packages was a huge success. Even the *Wall Street Journal*, rarely a commentator on packaging, took note of the bold new concept. The once regional brand has been the leading brand of ice cream throughout the U.S. ever since.

Aside from minor modifications, the original concept of Breyers ice cream packages, initially considered risky, has maintained its visionary appearance throughout the years. And Breyers continues to be a favorite brand of ice cream.

Had we, the designers, succumbed to the then traditional white dairy packaging, or hesitated in view of serious doubts on the client side, who knows where the brand would be today?

Or whether it would be at all.

Taking the initiative

What does this experience tell us?

The close working relationship between management and design agency, and management's willingness to listen to and seriously consider the designer's point of view, despite initial doubts, should serve as a shining example for smart business procedure; a guideline for everyone involved in a package design development program.

It spotlights the critical importance of *including* the package design agency as *a proactive partner*, participating throughout the strategy development process.

We realize that this may not become every marketer's normal procedure overnight. Habits die slowly.

Or, in the words of Mark Twain: "Habit is not to be flung out of the window by any man, but coaxed downstairs a step at the time."

There are many ways for designers to coax a client's confidence, step by step, on the way to becoming a proactive partner.

Becoming a proactive, visionary partner

Being in the business of servicing, it behoves design agencies to be proactive by monitoring the brands of their major clients, even *initiating* meetings to review brand strategies and float potential package design improvement for such brands, be they graphic, structural or technical.

"We have been thinking about the problem with your Brand X that you mentioned in our last meeting", the designer may say to their client. "We have a few ideas that we think might solve the problem. Could we come out to see you next week?"

What marketer would not be delighted with this kind of initiative by the design agency? This kind of proactive approach, involving a moderate investment by the design agency, would go a long way to reverse the clients frequently myopic perception of designers and allow them to see the design agency as a creative partner – a partner concerned about your business success, perhaps even your *personal* success.

Who cares if nine out of ten such initiatives will never get past the discussion stage. More importantly, it will spark new ideas and you will benefit from the creative input by the design agency and accept the design group as a concerned, proactive and visionary partner. A professional who is anticipating your company's needs and taking initiatives to help you to fulfill them.

It's a win–win situation for both you and your design consultant.

But – it takes two to tango

You cannot expect this to succeed if it is a one-sided relationship.

It is as important to understand and encourage a close working relationship with the design consultant as it is for the design consultant to understand your business sufficiently to be proactive.

Unless and until the design consultant is invited to sit around the strategy table together with you and your ad agency, from the beginning and throughout your strategic initiatives, the designer's ability to have an in-depth comprehension of your brand strategy will be limited. Thus you will deprive yourself of benefiting from the designer's visionary mind-set.

Pam DeCesare of Kraft Foods is an active proponent of in-depth working relationships with design agencies. She has a definite viewpoint on the subject and is not shy in expressing it:

> I see it as a real partnership, where the design firm is involved from the get-go, so that they understand intimately what the business is all about.

> What are the dynamics of the particular brand? What are we facing? What's the impetus to change or to grow? What are the long-range plans? And then, with all that information, you can develop the packaging, the timing, the cost, and what the brand wants to achieve.

> We want everybody involved. We want the ad agency involved. We want consumer promotions involved. We want public relations involved. We want consumer research involved and we want the design consulting agency involved. So a lot of people are really participating to provide their points of view, to provide their experience.

> When you take all these people on a little journey of insight and knowledge, the decision-making process becomes a lot more objective and a lot faster.

Matrix management

In the management jargon, 'matrix management' is what Ms. DeCesare is talking about. It is a system that has been activated in many major manufacturing companies to take advantage of the knowledge and experience of various functions that exist within the company. The system encourages individuals to participate in teams that share their knowledge in an effort to improve existing situations and initiate new, sometimes non-traditional methods of operation.

In a presentation to an audience of packaging professionals, Stan Zelesnik, director of education at the Institute of Packaging Professionals, explained:

> The traditional way used to be what we referred to as the packaging "value chain". All the ideas flow downward with a minimal upward flow. Ideas went from the raw material supplier to the converter then to the packager who, in turn, presented his ideas to the retailer and, finally, to the consumer.

> Here's how it has changed.

> Today all those key players in the packaging chain sit around a big table and communicate with each other. The material suppliers talk to retailers. Converters talk to consumers. Everyone talks with everyone else.

If that's the way new ideas are initiated within the chain of package engineering and package production within a company, would it not make equal sense – or even greater sense – to apply matrix management to the development of package design programs that would draw their strength from the combined input of all participants in the strategy development?

The designer at the strategy table

Consider this classic example: Tostitos

Tostitos, the well-known and highly successful brand marketed by Frito-Lay, was not initially conceived as the brand it is today. Originally, the product was meant to be a variety of Frito-Lay's popular Doritos snack line to exploit the success of Doritos by extending the line with a unique new flavor variety. But line extensions are not always a paragon. Some marketing experts believe that when you have a uniquely different product in a given category, treating it as a line extension of a well-known brand name is not always the wisest way to achieve success.

Frito-Lay's advertising agency at that time must have had this in mind when it discouraged the idea of treating the new flavor as a Doritos line extension. It saw the shape and flavor of the new product as being uniquely different in the chips category. It saw it as an opportunity for a unique new brand. Instead of crowding another flavor variety into the already multi-variety Doritos line, it recommended introducing the new chip as a *new brand,* with a new brand name, new brand identity, new advertising and new packaging. Frito-Lay marketing management enthusiastically approved this concept.

Olé, the Tostitos brand was born.

Frito-Lay's introduction of the Tostitos brand in a market already saturated with chips was fortified by this bold new package design

To achieve this strategically new position, Frito-Lay management and its agency did not go it alone. Everyone involved in the new venture, from marketing to product manufacture, advertising *and* package design, participated in the development of the new brand.

With everyone keenly aware of the impact of competitive snack packages at the point of sale, the packaging was a particularly critical component of the total brand concept. As the design consultants, we were invited to participate in every step of the Tostitos strategy development and to take part in every major meeting with Frito-Lay's marketing group.

Packaging concepts, advertising concepts and sales promotion concepts were thus reviewed with our client as an integrally conceived program. It was a classic example of effective matrix management.

And the rest is history. Tostitos has been one of Frito-Lay's premier success stories.

Chapter 17
Packaging politics

We think of politics as electioneering, new presidents, governors or mayors. In corporations, political changes are often made, bringing in new CEOs or new CFOs.

But there are also political aspects to package changes that take strong vision to overcome.

Packaging politics can take place within major package changes, with collaboration among packaging people, and even with the politics of digital commerce.

The political change

There are many companies that produce products and have been relying on their package suppliers for years to bring these products to market. Often the relationship has grown so that the package supplier's plant is adjacent to the producer in order to integrate the manufacturing and packaging function. This often develops into a symbiotic relationship between the manufacturer and the packager.

Let's take for example Pillsbury flour that has always been in paper bags. But if you go to a store, you'll notice how the paper bags often leak and the flour section can be a mess. These paper bags are also messy in the home.

Several years ago, Pillsbury secretly investigated the idea of using plastic bags to contain flour. The project was top secret because the paper bag supplier had been working closely with Pillsbury for many years, and this could mean jettisoning some of their business. Additionally, the Pillsbury plant and bag supplier were adjacent to one another. So a political issue became part of the functional packaging issue.

Different types of plastic and different weights of plastic were investigated, and it was found feasible to package flour in a plastic film, which would result in a package that was actually easier to ship, store, shelve and transport. And it created less of a mess at retail and at home. But the issue was too politically hot to handle. Suffice to say, the vision was there but the politics were not, and the plastic flour package never saw the light of day. It was given the sack.

You'll often see broken paper flour bags making a mess. But because of the commitment to the paper industry, don't expect to see plastic bags anytime soon

Political collaboration

The collaboration between the design consultant and the producer's in-house design staff can sometimes be delicate.

Several years ago, we were hired by Campbell's to redesign their line of Swanson frozen dinners. This was a major line of products requiring us to develop a whole program based on how the product was merchandised nationwide and how it would be positioned and branded for the future.

We had many meetings at our office and at the Campbell's offices in New Jersey and conducted research and store audits in different parts of the country. The audits provided information about consumer attitudes toward the brand as well as how the brand was merchandised in different regions and how it fared against competitive products.

We kept the client informed along the way and agreed on what the brand stands for and the potential approaches for changing the branding and graphics on its packaging. There was agreement on the brand architecture, showing how different products in the line related to the main brand, and also how line extensions would work with the main brand.

During these presentations, the key marketing people for Swanson were present along with some representatives from Campbell's in-house design department. After several weeks of this planning, we developed the design concepts and presented these ideas to the client at a large meeting. The presentation seemed to be a success!

Before final decisions were made for the next steps, the head of Campbell's in-house design department requested some time to make a presentation of its own. This was not on the agenda and had not been expected by us or by Campbell's marketing department. However, we all gave them the okay to make its presentation.

We were surprised. Its in-house design department brought in scores of presentation boards consisting of layouts, drawings and dummied-up packages to represent its ideas for the Swanson line of packages. Unbeknownst to us or the marketing department, it had been working for weeks on its own ideas, trying to outdo what we were doing.

The Campbell's design department ended its presentation thinking that it had overwhelmed and impressed the marketing department. However, it had done just the opposite. The top brass from Campbell's told it plainly that it had been wasting its time and that the Campbell's marketing department was shocked at that attitude of competition instead of collaboration. The design department was dismissed from the meeting and dismissed from the Swanson project.

So protecting its turf, and becoming possessive to the point of competition with Campbell's outside design consultants, had, until that time, kept Campbell's from moving to market quickly with visionary products.

Campbell's is a much more progressive company today, with better collaboration with its outside designers and their agencies. Campbell's now uses knowledgeable in-house design coordinators, working closely with outside design firms.

Digital politics

Digital commerce has its political drawbacks. Because of the Internet, many reputable products in the drug industry are taken advantage of.

For example, let's take a product from a major pharmaceutical company, which produces a medication and sells it for 100% of its price in the U.S. market. However, it will sell it to a country with a smaller economy and less wealth at a lower price – perhaps, at 50 cents on the dollar.

That same product may then go back on the Internet and be sold back to the U.S. market at 75%, taking a 25% profit. This isn't legal, but it's done because the Internet has made the world a place where pricing can be gathered from many different sources. This works like a gray market and the challenge will be to see that those products sold at a discount cannot get back into a market where there is no discount.

This type of digital politics can be solved through the emerging software specialists who provide package management systems, where the complex permutations of package copy used for the different countries can be easy to manage and implement in order to help protect against parallel importing. According to J.P. Terry, CEO of BrandWizard Technologies in New York:

> We address the issue of parallel importing in a number of ways including special bar coding, special regional graphics and pre-qualifying the printers, so that the printers will use special printing techniques. It requires a sophisticated system such as our Packaging Wizard to ensure all these variables are applied automatically. Without such systems, these measures can place huge demands on package production studios.

The special printing techniques are particularly interesting because they are not visible to the re-importers, but are detectable by customs agents. A special chemical is added to the printing inks that are used to print the carton and label graphics. This chemical is not visible to the eye but detectable by a special optical scanner. This is one of the most successful methods of curbing counterfeiters and re-importers.

Chapter 18
Virtual package management

The only thing that saves money is efficiency, and package goods companies have rarely won awards for efficient marketing departments.

Several years ago a company's marketing department could create a brand and promote and disseminate its information through printing suppliers and advertising agencies.

It's not that simple today. With the proliferation of branded products, and the speed needed to get to market, and the need for the brand information to be disseminated nationally and very often worldwide, you can't do it the old way.

So how did manufacturers try to become more efficient? They started putting their brand information on their websites and emailing their packaging and brand information to printers, design agencies, ad agencies, retailers or anyone who asked for their products. This handcrafted information turned into an impossible enterprise for a lot of companies, with low efficiency, high cost and substantial production errors.

How was this solved? A new industry was formed to focus on creating value by leveraging technology for package goods operations. The industry is based on outside vendors who, through web-based technology, manage the brand components regionally, nationally or worldwide for package goods companies.

What is this industry called?

Although this book calls it 'virtual management', there are hundreds of different names used to define it. Gartner, a leading technology analyst, calls it 'marketing resource management', some call it 'integration management', some call it 'web-based knowledge management' and some call it 'digital asset management'.

But it all adds up to virtual management – and package goods companies can use it as a way to get information out to the world more cost-effectively, accurately and efficiently.

When do you need virtual management?

Almost every package goods company needs it in one way or another.

If a company is introducing a brand new product and needs to gather information about the product to inform customers, its sales force, along with its own finance, legal, marketing, sales, manufacturing, channel management, this company needs to find an efficient way to get out the information.

If a company has hundreds or thousands of packaged products, it needs an efficient way not only to get the brand information to all the constituents, but also to update this information quickly, easily and cost-effectively on an ongoing basis.

The three systems

Which virtual management system is best for you?

They each have a different specialty, and each marketing department or package goods company has to decide which is the best fit. With some companies, they may need more than one virtual management system in place.

For this book, we'll break them down into three types of systems. Each system is an offshoot of a marketing resource that has existed for years:

1. *Offshoot of the product catalog* that used to disseminate information about products and brands to retailers and the printers and agencies associated with these retailers.
2. *Offshoot of the converter or printer* whose main customer was the package goods company.
3. *Offshoot of the design and branding agency* who provided consulting and art work for their package goods clients, which was used by converters and printers for their packaging and advertising.

Offshooting the catalog

Go to a large retailer's website, let's say Target, and you'll start to get the picture. These retailers are maintaining an aggregated catalog with information from thousands of suppliers. Look at all the brand information and the product description they maintain. And the odds are that each of those suppliers is giving information to all the other retailers.

So companies like HAHT Commerce, according to Rowland Archer, its chief technology officer and co-founder, help the package goods companies and the retailers to look at their product information and brand information more systematically by creating a repository inside their company of the facts that their customers need. Mr. Archer calls this "one version of the truth".

With your new product being launched, the marketing guys create their descriptions for marketing text and get some photographs together. The finance guys hammer out the price. The manufacturing guys are figuring the size of the carton to ship. And that information has to be quickly sent to a retailer partner and your salespeople in the field. So you need a catalog or repository inside your company.

Archer's company provides software for its customers, like Pfizer, which maintains its product information. If Wal-Mart, for example, as one of Pfizer's customers, wants to get information, it retrieves it from Pfizer's digital catalog. And this is the way information can be exchanged worldwide.

Now, how can you set standards for how the information is shared? A group called UCC.net has developed 62 standardized attributes, like height, weight, color – basic information – and eventually this will expand to 151 attributes. So the retailers are telling the manufacturers, "We want to know everything there is to know about your product that will help us sell more of it."

The vision for this type of enterprise is eventually to have retailers using the same descriptions of their products. Although these descriptions will describe the product in a certain order, it won't affect the branding. Coca-Cola would use messaging for its products different to that from Pepsi, although from a catalog order standpoint, they might be described similarly.

Where are the opportunities in this?

Because of the nature of business, there is often a desire to give preferential information. So this dissemination of information will give both the packager and the retailer more control over what information is disseminated, when and to whom.

This also will help packaging and retailing through the opportunity of more quickly modifying a package based on consumer buying patterns.

A music company selling DVDs through Wal-Mart, Blockbuster and Target will be able to track how those DVDs are selling, whether they should be bundled together, whether they should have special packs by the cash register, and get the sales data on a timely basis. The company will then be able to analyze it and decide how to change things, perhaps in outlets in different regions.

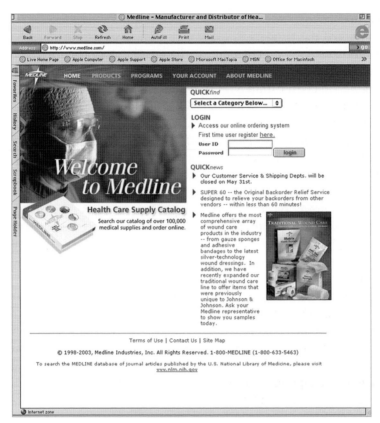

*A company like Medline provides product information
for its customers through its website. It manages the
product data with software from Haht Commerce*

So if it's selling through Wal-Mart, Target and Blockbuster in a five-mile radius, it can see what's really selling well in these stores, and use that information to adjust the in-store displays.

Approvals up the line

Package suppliers now have standard processes for getting things created and approved. Digital management tools allow them to institutionalize the process and notify their retailer customers once a new product has been approved. A group called GCI, or Global Commerce Initiative, is a global umbrella organization setting standards for consumer package goods. If someone in France wants to see a particular product and get information on that product, he or she can look it up in the virtual catalog, which would also refer to how it would be marketed in France.

Extra offerings

In the past, there was a limitation on how much information retailers could effectively manage. The more information they had, the more people they needed to keep it up to date. The new technology helps manage the data from the packager, and the retailer doesn't have to worry as much about keeping it up to date, getting new pictures of the product or getting new information. The package goods companies take care of that and the retailer can get it as soon as it is updated.

So package goods companies can offer retailers things they would never actually print on packaging. For example, they could have beauty tips, or describe complementary products. And in time they'll even have things like video. With a cosmetic product, they could show someone applying the makeup.

The amount of information is richer through the virtual catalog and gives a better brand experience to the consumer who's shopping for it. In some cases, the product information could be at the producer's website as well as the retailer's website.

There will be a period of investment over the next few years, and this technology will allow the brand experience to be more tightly controlled and personalized than in the past. Package goods companies with vision are adapting these new methods of cataloging their product lines, and the retailers will love it.

Offshooting the printer

How has the printing industry changed for package goods companies? This industry is also focusing on creating value by leveraging technology in its operations. Instead of serving as converters or printers for the large package goods corporations, many in the printing industry have introduced digital imaging solutions:

- They place people and equipment at their customers' locations.
- They provide systems for their customers to track and retrieve the customers' digital assets.
- They have access to and sometimes own design agencies and production art centers.
- They provide pre-press, color management, digital imaging and print management for their customers.

They usually serve the packaging and advertising market through their knowledge in pre-press, imaging and production.

But are they creative?

Some of these companies, such as Schawk or Southern Graphics Systems, have linked themselves with design consultants who help them provide creative solutions. But their specialities are still digital imaging and management solutions for the digital pre-press imaging market, along with providing color and information in a digital format.

How has this changed packaging?

You could say it's one-stop shopping from the origination of a package to its production. However, no one can be an expert at everything that encompasses the package spectrum.

The vision that this part of the industry had in placing people and equipment on site at client locations, and using technology to manage the information from these various locations, is their strongest point. However, the virtual nature of the web will continue to place pressure on the print industry, since what was previously a manual process is today becoming automated.

And by assembling design, pre-press and display companies under the printing banner, they are a visionary part of packaging's future.

The offshoot of the brand consultant

"Get to market more quickly!"

These are today's buzz-words, along with "Keep the costs down!"

Speed, economy and, let's add, brand consistency. This is the third virtual management focus on value by leveraging technology.

According to J.P. Terry, CEO of BrandWizard Technologies, many industries that rely on packages to deliver their products to the market and to the consumer have these important requirements – speed, economy and consistency.

Take for example, the pharmaceutical companies. In the U.S., once they get FDA approval on a product, guess how quickly they want that product on the retail shelf? Seventy-two hours later. Very tight timelines!

BrandWizard develops systems which deliver these efficiencies, by saying, "Don't take the 100 manual steps to develop this package. Let's let the user focus on the 10 steps that are important and *automate* the 90 repetitive steps." This approach provides a new paradigm for creating artwork, especially for streamlining the development of packaging.

It's based on a brand system that becomes the overall theme for all the packages. Pharmaceutical companies, and companies in many other industries, need and use a consistent branding system. This doesn't mean that the packages look just like one another, but they will have an overall brand theme, as pharmaceutical brands usually have.

So it's not about one package or one label, it's about the entire group of packages and labels. And the computer can be programmed to provide the hundreds of thousands of elements that become the artwork for the hundreds or thousands of sizes, shapes, logos, illustrations and such that make up the brand.

A company like HP, with thousands of different SKU's, saves valuable time with BrandWizard software automatically assembling its artwork elements

So how do you benefit from the system?

Let's assume the brand system has been designed and has a visual theme that carries through on all the packaging. Using BrandWizard means not having to create the artwork elements by hand anymore. The elements for the brand are in the system, and the user goes step by step down the list, making a series of choices – measurements, brand signature, trademark and so forth – and the software automatically assembles all these elements into the full package artwork, which is ready for approval and sending to the printer.

In fact, the process of creating the package graphic is done in only minutes. And within a day, it can include the artwork, proof reading, and approval among marketing and legal departments – a proactive approach to packaging.

As a vision for global packaging, instead of finding out that Zimbabwe has made an error in the trademark two months later, Zimbabwe gets it right the first time. This shatters conventional thinking, which, in global packaging, was more reactive than proactive in the past.

So the computer is enforcing the proper implementation of the package, giving better consistency, ensuring its trademarks, and doing it in much less time.

Let's start big. General Motors has hundreds of thousands of different packages that are used for the parts that it supplies to dealers and parts departments nationwide. It was spending into the millions of dollars annually for packaging artwork preparation. With Brand-Wizard Technologies, they no longer use an art studio. Inside General Motors, a small staff of people put the information into the Brand-Wizard program and it's automatically created. Companies like Hewlett-Packard and Becton Dickinson do the same.

Will this make the in-house designer extinct?

"If you use a digital management system, I'm quitting," said in-house packaging graphic designers within many companies.

But once the technology is brought into the company, that graphic designer has become more valuable, because digital management tools have taken the repetitive manual tasks out of the hands of the production artist, leaving more time for designing.

That in-house designer can now upload master artwork, set up templates, and put in all the packaging information so that everyone worldwide is getting the brand information consistently. This has changed their job description in a positive way.

It means that a brand within many different countries can be handled efficiently and consistently – certainly a change in the paradigm for global packaging. Drug companies can reduce and rationalize the quantity of different sizes and shapes. This visionary approach makes their packaging streamlined and successful.

Can you virtually manage advertising?

The same type of program used in packaging can also be used for advertising.

For example, after September 11, General Motors switched over to a "Keep America Rolling" campaign. Through information technology, General Motors' ad agencies, D'Arcy and Campbell Ewald, were able to create new ad templates, new master templates, and then upload them into the BrandWizard system. General Motors' 7,800 dealers were already logged on and the ads were out that weekend.

Remaining focused

Virtual management companies need to remain focused on what they do best. Their vision should be to combine with others seamlessly when it means integrating work flow systems, corporate databases and corporate systems.

There's a lot of competition today with digital asset management. This area will grow as more and more corporations realize that they can get speed to market and general efficiencies through the digital packaging methods.

Some companies will want all their services taken off their hands, from design, artwork, digital management, work flow, to pre-press and final printing and fabrication.

Other companies are reluctant to give all those services to outsiders and would prefer service organizations focused on one or some of those areas.

How can you get out of it?

You have turned over your digital assets for virtual management by service providers. Some feel that it's like handing over a child for adoption, because in the past, service providers have profited from the difficulty that clients have in getting out of their system.

Think of the concern that the person within the corporation might have because an outsider is putting their information into the outsider's system.

How do you mitigate this risk? You should ask beforehand how to get out of the system if you have a disagreement in one, two, three or four years. You'll need to know how you get your stuff back – your logos, your laserjet pictures.

The smart service providers with a vision of the future will develop their information and databases so that you can get out of the system if, for some reason, you need to. This is important, and corporations signing up for digital management need to make sure they can reclaim their assets when they want them.

The virtual management of packaging and advertising has led to efficiencies and value creation by leveraging technology and is the way of the future.

So these virtual management businesses will grow in the future, and their opportunity is in giving the best service and focusing on specific needs that are important to marketers, package goods companies and retailers.

Chapter 19
All you need to know

The categories blur

There is an increasing blur between drugs, foods and dietary supplements.

You can find products side by side on the retail shelf that appear to be similar to one another, yet they're regulated differently because one is a drug and one is a supplement.

Or one may be a conventional food and one a supplement.

So you end up with similar things. Sooner or later, the producer of the product that has the tougher regulatory approval wonders why he's going to all this trouble to get clearance. He's competing with a dietary supplement that doesn't have to go through that clearance trouble. Today you find yourself with different regulations depending on the kind of product that you have.

In fact there are prescription drugs, over-the-counter drugs, dietary supplements and traditional foods, and you could come up with examples where you have one product from each category aiming to do the same thing for you. Lowering your cholesterol probably falls into products from all four of those categories.

Now, the drugs have to be cleared by FDA, but the dietary supplements do not. And the foods are on the same basis as the supplements. So each one has a different regulatory task, although its products all have the same benefit.

What can we do? We have to see where the regulations do fall and stay tuned in so that packaging can conform without losing its creativity. This keeps the marketers and the designers busy.

Freedom of speech

From the other end of the information spectrum, there is a cry for less regulation. "The concept of freedom of speech may now be open with labeling issues", comments Eric Greenberg, an attorney in Chicago, who has given many speeches and seminars on the legal aspects about packaging. He suspects some legal requirements might change.

The drug industry succeeded in getting more freedom in making claims for drugs. As Eric Greenberg indicates, you don't start from the premise that the FDA can do whatever it wants and then make sure it's reasonable. You start from the other premise, that the FDA cannot do anything without good reason, and then you can see if what it does is really necessary.

Could you imagine that in the next few years, by keeping with tradition and the constitutional requirement of free speech, anyone can do what they want on a package? Unless, of course, it's misleading and unless the rule is absolutely necessary for the purposes of protecting public health.

So today there is a possibility of a major change in the bedrock principles of how speech is regulated on food and drug labels and in advertising as well.

Couldn't you take advantage of this so-called free speech and have things on your packages that are misinterpreted by the public, or exaggerated to the public? Because if freer rein was given, one result could be information which is false or misleading to the consumer.

So you have to look at specific requirements that are being questioned, and try to figure out which ones would change.

For example, health claims may not have to be specifically cleared by the FDA, but everyone may be expected to put nutritional panels on foods in keeping with existing regulations. If the FDA said it will not allow the manufacturer to put anything on a food label about the food's effect on a disease, unless it becomes a medical consensus that the food really prevents or helps treat the disease, you might say that's logical.

But marketers could say, "Wait, I have a well-designed, well respected, statistically significant study that suggests that my product, when correctly applied, might encourage hair growth. Why can't I say that on my label? I won't mislead people and say that it's been proven, and I'm not going say that the FDA agrees, but I should be able to tell the public about the existence of this study."

So in the future, marketers might not be prohibited from mentioning the study and, with appropriate caveats, qualifications and limits, people would hopefully not be misled about the information being conveyed.

Warning! Dangerous!

There are packages that require a lot of warning information. For example, pesticide products used at home.

These will change in the future. With RFIDs and computer chips, one could eventually embed on the side of a can of bug spray a button with the recorded message verbally giving a warning. For products with a lot of label information and concern for safety, this is a good approach for the future. The most important warning information in this recorded message plays by pushing the button, and the message could fit into ten languages if it's on this computer chip.

So where are we going with legal information requirements?

Well, perhaps free speech will change some creative aspects of the information one sees on a package, and computer chips may help make our products even safer.

Why is all this package information important?

Well, where is the sales clerk? In the large retail environments, the package has to work harder than it's ever worked before. It has to tell the whole story, since there is very little sales component any more. The stores talk about good customer service, and especially show them in their advertising, but do we really see them?

So the packaging has to teach the consumer about the product, and often teaches the retail sales clerk about the product as well. It speaks for the brand and speaks for the product because no one else will be doing it. It is the last step of the pathway to the purchase. Remember, the package is the final "go" or "no go" decision maker.

section five

the future of
visionary
packaging

Chapter 20
Package opportunities and challenges

An old saying goes: "If you want things to stay as they are, things will have to change."

Think how businesses have changed with new tools, such as email, the Internet, computer chips, hand-held PDAs, and many other things that did not exist even a few years ago. This brings us many new challenges and opportunities that we never had before.

A big opportunity is with innovative product packaging. This was made especially clear in a speech made at a business conference by C. Manley Mopus.

Where are these opportunities for packaging and branding? According to Mr. Mopus, the focus has shifted away from viewing women as the cooks and caretakers of the family. Let's think about the many products geared specifically to women's own needs in health foods, cereals, and even many auto products. All these categories and industries that have been targeting their packaging design efforts toward males now have the opportunity to shift focus.

The food opportunity

Have you noticed the growth of the frozen food category? The packaging implications of frozen food are tremendous.

Wal-Mart and some other chains may have 60 doors of frozen food competing for attention. Vast competition, with many of the producers overlapping similar products.

For example, each producer has its own macaroni-and-cheese, spaghetti, glazed chicken and so on. In fact, if you only had one producer with all the different frozen food products, there would probably be four doors instead of 60 in the frozen food department.

The growth of the frozen food category opens opportunities for new products, while currently the similar products vie for attention

Another opportunity is with the dinner trays themselves within the frozen packages. The heated dinner trays contain different foods within different compartments. Yet we cook all the components together in our oven or microwave. One of the components may taste great, but what about the other components in the other compartments?

So a big opportunity will be to develop a cooking tray for frozen food that compartmentalizes the different products, so that even though they are cooked or heated for the same amount of time, their temperature is appropriate for each item. It will be interesting to see which producer will do this.

Research brings opportunities

There is nothing permanent except change. *Heraclitus*

Some say that if we had used research for car design 60 years ago, cars would still all be black. But research methods have changed considerably, and research has become a powerful tool in determining how much change and how far a brand can go to change its package and its image. When change is needed, we don't want to miss out on the opportunity.

Consumer research in packaging design has fundamentally changed the way decisions are made for package changes and introducing new packages. Since some marketers only make a package change when they know that it will be accepted by the consumer, today's consumer becomes part of the design or selection process.

What kind of research should be used? According to Elliot Young, chairman of Perception Research Services, the combination of qualitative and quantitative research is important. With qualitative you listen to people, and even if the discussion digresses and the people discuss a subject that isn't anticipated but is relevant, you let them discuss it. Qualitative research, usually in the form of focus groups, offers much flexibility and can provide you with reassurance that you're not confusing or offending your consumer. Its drawback is its inability to provide numerical evidence for guiding your decisions.

The quantitative research can later give you actual numbers. With quantitative research, or survey research, feedback is gathered from hundreds of consumers through structured surveys and rating scales to guide decision making. Survey research can measure package performance for shelf presence, esthetic appeal, shelf impact, product expectations, brand imagery, purchase interest and so on.

And online research is fast and can get answers regarding brand recognition and brand name. But when online research applies to packaging, experts must handle it.

Research today shows that companies are taking more risks and looking for more opportunities. Top management people are under pressure to generate positive numbers. So what do they do? First, they cut staff and overheads. Second, they try product innovations and try putting life into their big brands. This should be reversed! The product innovations and brand building should be the start of the road to generate positive numbers.

Fortunately, the risks of product innovation can be mitigated through consumer research, so it has become a popular method for making business decisions.

But can survey research, or simulated shopping, accurately gauge the sales impact of new packaging? In reality, no matter how realistic and complex researchers make their shopping experiences and sales models, accurate future sales diagnostics are uncertain. The research can identify the changes from current packaging which are likely to improve sales, uncover risks associated with making package changes, and project whether new packaging is likely to have a positive or negative effect at retail.

Who is making the changes within the package goods companies?

Believe it or not, it seems like the heads of companies are often more willing to make changes than the middle managers. Elliot Young has found through research that many middle managers are petrified that they'll hurt the brand by making changes.

When is the opportunity for change? Many times, changes are made because the heads of companies come from other companies and want to make their mark in the new company. They're not as much concerned about the heritage of the brand, and in many cases they're right.

When there is a real benefit to a package change, the consumer buys into it. These opportunities can be researched before the changes are made. You can't describe the changes verbally, but can show them to the consumer. Let them touch it, feel it and describe it.

The fresh meat opportunity

The grocery store in the future needs no butchers.

This is a big opportunity for the meat and poultry industry. Packaging will contain meat ready for the display case. Individual retail packing will be done at a central location, and then distributed to the grocery store, and the store clerk will only need to put it on display.

This makes for big cost savings and enables control and centrality. A good example of this is the Jennie-O product line, including ground turkey, in easy-to-read and appealing packages.

Strong branding, good copy and appetizing photography have helped the success of the pre-cooked Jenny-O line

Notice also how organic meats are starting to appear with specific brand names.

If meat is kept within modified atmosphere packaging, it can stay fresh for three weeks. The problem is that it turns from red to brown during that time, but is not unwholesome.

So where is your opportunity? Well, many companies are trying to develop a package that keeps meat fresh for all those days, but still keeps the meat looking good. As it is important to keep it safe, it is also important to keep it looking and smelling good. Therefore, a system that holds the meat's appearance, as well as maintaining wholesomeness will be a welcome innovation.

The natural instinct

Where is the opportunity for private label brands to dominate and differentiate themselves from the others? You will notice that there is no real brand dominator with natural and organic foods. What an opportunity for private label brands, since it's still available territory. U.S. retailers seem to be behind the trend with organic foods and should take advantage of the opportunity to create ownable territories for themselves.

It's been shown that today's consumer has an increasing sensitivity toward genetically modified ingredients, chemicals and pesticides in products. So a huge trend toward organic should follow in the future. As marketers, let's not lose that opportunity.

The challenges you face

Are you concerned that within a few years there will only be a few mass retailers worldwide, such as Wal-Mart, Costco, Target and a few others? How are these mass retailers, who sold general merchandise through the years, able to compete with one another? These merchandising giants look for ways to outdo the other.

The mass retailers and the smaller retailers all realize that shopping for food brings people into their store. We've all noticed how some stores that at one time had no food products now sell an abundance of food products to attract the shopper.

Wal-Mart builds more Super Centers, and Target more Super Stores and they add many food items to their already popular non-food items. In fact, Wal-Mart has almost $60 billion yearly in food sales.

This makes you face many future challenges – the mega-stores, the merger mania, the proliferation of brands and product information, pilferage, growth of the senior market, legal issues and brand strategy. So how can branding and packaging help retailers and producers to meet these challenges? Let's discuss some of the challenges.

The merger mania

When two store chains merge, where does the brand loyalty go?

With retail consolidation through mergers and acquisitions running at an unusual pace, a big challenge will be to maintain brand loyalty among consumers, especially with their buying habits at these retail arenas. It will be especially important for retailers to give a pleasurable shopping experience to those consumers who bought at either or both of the merged brand enterprises.

Retail consolidation often means redesigned packages as well as redesigned stores, signage, trucks and many other visual media. You must be especially careful to develop a design program that understands the consumer of both of the merged parties, and understands the vision for the future of the retailer to appeal to those consumers.

With good planning and designing, the consumer will feel positive about the merger of the companies, or the acquisition of the brands, as long as the level of consumer service, product quality and brand identification remains strong.

There is no doubt that the experience of shopping for well-placed packages on shelves and good brand identification and information increases consumer satisfaction with the shopping experience.

How many brands do we need?

Let's call it the challenge of the crowded shelf. With shelves getting more crowded, the criteria are much greater to ensure success for new line extensions and new initiatives. The standards are higher for putting products out at retail. And in the future, retailers may start demanding their suppliers to start whittling down the options. This could change the way packages are designed in the future.

If many products become part of single mega-brands, it will mean that some of the products that you now have as individual brands will be moved into a mega-brand with a different name. Successfully growing these mega-brands will depend upon your vision in representing them on packaging. This is a big challenge and opportunity for designers.

At companies like Nestlé and Kraft Foods, where they are constantly launching new line extensions and new technologies, it's important to know how to place the product in the marketing mix.

Think of the large package goods companies that need to meet this challenge. Take Procter & Gamble alone. That company has over 300 brands, mostly sold in mass-merchandising, grocery and drug stores and chains.

So what is the right architecture for a brand? How do you divide up the products and packages in a mega-brand, so that your consumer understands the value you offer them? And when should it be just an entirely separate brand? Brand architecture provides a structure for managing and nurturing the range of your company's brands, products and services. You can organize the products and packages for your target consumer, and express the value offer to the consumer who buys the products.

There are four basic brand architecture options: *masterbrand, overbrand, endorsement* and *freestanding*.

The *masterbrand* option is appropriate when consumers make buying decisions based upon the producer's brand name, and when funds for brand building may be constrained. For example, Rubbermaid is a masterbrand under which there are thousands of products in various household categories.

The *overbrand* is appropriate when the producer's brand name is instrumental, but can be enhanced by association with key products. Kellogg's is an example of an overbrand, and the Kellogg's name is normally given less emphasis on its packages than the product brand names, such as Frosted Flakes, Pop-tarts, and NutriGrain.

The *endorsed* brand, or brand endorsement, is used when the focus is on the product level, and the role of the producer in the value proposition is minimal. An example of endorsed branding is Nabisco, which uses its brand name at the top corner of all its products, but greatly subordinates the Nabisco brand to the product brands such as Oreo, Teddy Grahams and Chips Ahoy!

The *freestanding* brand option is appropriate when each of the producers' individual brands offers unique value propositions for distinct, targeted consumers. Although Procter & Gamble is the producer of many products, its brands are freestanding, and identify Procter & Gamble only as a signature on the back or side package panels. As the name implies, the freestanding brands such as Crest, Scope, Cheer, Pringles and many others stand on their own.

With the introduction of any product, you need vision and strategy to decide on how the product fits within your brand architecture. Some of the decisions you need to make for a branded product can be seen on the brand strategy decision tree below.

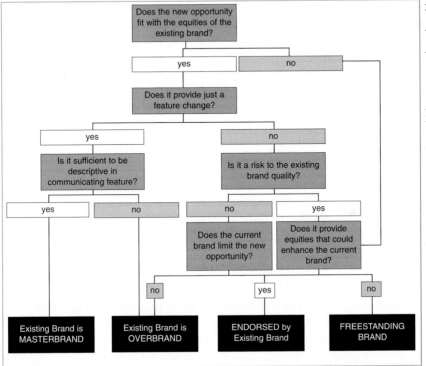

Following the route in this chart that best suits your new product introduction can help you determine which option to use for your brand name: masterbrand, overbrand, endorsement or freestanding

As brands proliferate and packaging competes for attention, it also becomes a real challenge to keep packages looking good.

It's new! It's different!

"Violators" – flags or banners used to get your attention – are used more strongly than in the past to show when things are new. Of course, this creates more noise and visual competition at retail. So this challenge is in keeping the package simple and uncluttered. This is especially true for brands that have been elevated to look more like prestige products. There's a word to describe this, "masstige," meaning mass-produced products that look like prestige products and try to emulate the look that, say, Lancôme might convey with its high end skin care products.

How do you drive awareness and gain trial for a new product? Often we'll see another product attached to the new product by a shrink wrap in order to drive trial. And of course, the flags or violators that have been used for years are enlarged now to show when things are new, creating more clutter. So the big challenge is to simplify things and break through the clutter in order to gain trial for the new product.

Thwarting thieves

Theft and pilferage has become a major issue with some products, such as Gillette's razors and other smaller items that go for hefty prices. Companies like Gillette have risen to the challenge by offering structural features that often make the package harder to open in the store or conceal.

This is a double-edged sword. There's always the dichotomy of making a package hard to open to prevent pilferage, but yet not so hard to open that when you get it home it becomes frustrating. In fact, Gillette receives complaints from consumers that do find it hard to get into some of its packages. Gillette takes great pains to make the packaging easier for the consumer to open once they take the time to look closely at the package, but it's harder for a thief who has to quickly open the packages and get the product in his or her pocket before being spotted.

Tamper evidence is becoming more necessary and a challenge in our terror-oriented world. Among the early terrorists to invade our everyday lives we saw the episode with Tylenol several years ago. This was solved through tamper-evident packaging, which has become prevalent, not only in the drug category, but throughout every category of food. For example, many salad dressings, sauces and condiments have wraps overlapping their closures. It's even used in hardware stores to prevent pilferage. But let's not go overboard. The consumer still has to open the package and get to the product.

Many items now are packaged in clear plastic with sensors that have to be scanned, in order to help reduce the huge losses that retailers suffer due to pilferage, especially with expensive items and electronic items that stores can't always keep under lock and key.

Do older people use more packaging?

We know that as people mature they spend more on certain categories of products, especially health and medicinal products. The senior market is becoming a much larger slice of our population. One of the biggest challenges in branding and packaging will be meeting the needs of this growing senior market.

Solutions sound simple, but it takes vision to implement them:

- Striking the right balance between opening a bottle and keeping it child-resistant.
- Using larger type, but not crowding the package. You need better copywriters to make the messages shorter and more succinct.
- Using color and warning messages that don't turn off other consumers. Striking the right balance for a strong, but not morbid, warning.

These can be solved, and these are your opportunities and your challenges. It's critical that these challenges are met.

Brian Perkins, worldwide chairman Consumer Pharmaceuticals and Nutritional Group at J&J, indicated to us that another revolution is coming to the packaging world. He notes that within the next few years, baby boomers will stop tolerating current packages that are difficult to open, close and carry.

These people fight the idea of aging, and they want to stay young forever. But they won't be able to fight the arthritis that occurs in their hands.

They'll complain that manufacturers need to make it easier to open packages.

These baby boomers want to look good, want to appear fit, but they won't be willing to trade off convenience. The packages of the future will have to be made convenient for them.

This will affect the readability of the package panels, their inserts and, of course, will affect many product categories, not just the drug category.

The drug challenge

Prescription drugs today are heavily advertised. After a few years, many are sold over the counter.

But packaging and brand identity starts long before the product switches to over the counter. According to Brian Perkins, it's really important to get into the brand's DNA – the core or the meaning of the brand – so that when it goes from prescription to over the counter, the physicians, pharmacists and consumers still follow it, even though the name and the packaging may often be changed.

Another challenge is that the demand for legal information is going to increase on drug packaging. There will be more government information, more words, more print, more package inserts, and more information to the consumer. The government is requiring these changes and packaging will have to accommodate them.

And the packaging requirements or changes won't be radically different in different parts of the world.

Don't imitate

There is a challenge for creativity. Many designers are stymied by package regulations, especially in the U.S. Creatively, they'd like to be able to move in more innovative ways.

Let's compare the challenge of creativity to the automobile. An automobile always needs wheels, headlights in a certain position and, of course, doors and windows. You need room for passengers, the engine, mechanical components and cargo. But we see how cars have become very creative – perhaps not fully comfortable at all times. So even when designing a package, there's still room for creativity, but the challenge is to meet the functional requirements.

Of course, many lawyers and government regulators will tell you that the primary purpose of the package is to protect and preserve the product for safety reasons. From their point of view, the appearance and brand identity, or its communication, are secondary.

If you're creative, you can meet this challenge and find ways to position legal package elements, such as nutritional requirements on food packaging, so that the packages still reflect the promise of the brand.

> When people are free to do as they please, they usually imitate each other.
>
> *Eric Hoffer*

A big challenge is overcoming the imitation of package designs, legally known as 'trade dress'. One sees much imitation at retail, and often store brands imitate national brands. With all the possibilities for brands to stand on their own and gain memorability, there's no reason to imitate.

When a well-known brand believes its trade dress is being imitated, it will often issue a strong legal statement to the competitor. So the challenge is really to make sure that brands look different and can work on their own.

Store brands that imitate the national brand sometimes get away with it. In-house attorneys and personnel at the big retail chains often believe that they have a symbiotic thing going with their product. The producer wants shelf space in their store which represents a huge amount of their sales, so the producer tolerates a little bit of imitating by the store brand.

Listerine is an example of a major brand with many store brand imitators. Its bottle shape, with relatively sharp angles, sits next to near-identical copies in several drug chains. Although Pfizer, Listerine's parent, has managed to keep accurate copycats off the shelf, the copies are close enough to draw in some consumers. Which makes it more urgent and important for Listerine, or any brand that's threatened by copycats, to emphasize their brand name and promote the brand consistently for brand loyalty.

Of course, if the retail chain was a complete stranger, they would be asked to cease and desist. But when they're putting the producer's brand on the shelf of 2,500 of these stores, the producer often keeps silent.

Retailers can't take it for granted, though, because many large producers evaluate their potential losses due to imitation. There are plenty of lawsuits currently on the docket reflecting disagreements about the imitation of the brand packaging.

But remember, loyalty to your product is best built through vision and uniqueness.

Walgreen's store brand Wal-Tussin borrows part of the name and the distinctive green color from the category leader. Some consumers may be fooled

The consultant opportunity

There are tiers of brands. Even tiers within value and discount brands.

In fact, stores like Sears and JC Penney have had a hard time because they've aimed for the middle income consumer, where other stores, such as Target, have said: "Okay, we'll sell on low price." But along with that, Target, will take its branding and bring in a consumer shopping experience and a feeling that the packaging and the products are quality and value within this low price range.

So many retailers have lost touch with the outside world's perception of their brands. Although they think they understand the brand, they often need an outsider's perspective to do some research and discover what consumers really perceive about it:

- Where does my brand fit with the competition?
- What is its point of difference?
- Where can I take my brand strategy to fit into where I want to take my brand?
- How can packaging, merchandising, advertising and marketing reinforce my brand imagery?

So the opportunity is there for brand consultancies and designers to knock on the doors of the many retailers with these thoughts, and present them with the opportunity to invigorate their brands.

Chapter 21
What's ahead?

> The art of prophecy is difficult, especially with regard to the future.

This quote by Mark Twain is as true today as it was when it was first spoken many years ago.

But even if making prophecies about the future of retail marketing may seem like an exercise of futility, there are enough indications of where retail commerce is moving to be able to predict some of the directions that packaging will take in the future.

We have already discussed the revolution that will influence the marketing of brands substantially in the next few years. The meteoric growth of mass-merchandising outlets, such as Wal-Mart, Target, Home Depot, Cosco and others, has been creating an upheaval in the retail world.

But upheavals are nothing new. What is happing in the retail business today mirrors the same type of convulsion that mom-and-pop groceries experienced when supermarkets began to take over the grocery business a few decades ago.

As mega-merchandisers continue to proliferate, manufacturers and marketers will need, as we suggested, a more precisely targeted approach to package design as part of their overall marketing strategies, because the shortage of knowledgeable store personnel demands closer attention to package design.

As we said, the store clerk in many hypermarkets is disappearing and is replaced by packaging that must be uniquely qualified not only to attract the attention of the shopper but also to convey information about products concisely and effectively.

This is not an easy task. It requires that the marketers have a clear vision of how to communicate to the shopper what is important and beneficial about the product. In turn, it requires the designer to clearly understand how to use this information to create a visually and functionally compelling package that can propel the product into a sale.

The provocative approach of the electronics industry is an example of how entertaining consumers through the constant introduction of new products, or the reintroduction of improved products, can lead to exciting growth potentials. Using unique and visionary packaging is a particularly critical component of creating sales incentives at the point of sale.

Let the imagination soar

For better or worse, we live in a world of tumultuous activity. Hyper-outlets, wireless communications, e-commerce and the ever more mobile mass of consumers all contribute to our current way of life of turbulence and agitation.

In this frenzy of activities, the role that visionary packaging will play in branding, conveying product information and providing product enhancement is bound to be one of the keys to future product success.

This calls for marketers and designers to let their imagination soar to achieve new, unique and visionary concepts in both product development and the packaging of these products. To be sure, using unique and visionary packaging may demand some risks. But risks will be the way of life in the commercial world of tomorrow.

Yet, along with the risks comes the responsibility of understanding external issues that will influence package design directions and decisions.

The social issues of global marketing

Looking beyond our current understanding of the process of selling products, packaging in the future will need to broaden its criteria. Disregarding, for the moment, the dramatic influence of marketing retail products at hypermarkets and on the Internet, one of the most important adjustments – and possibly one of the most neglected at this point of time – is the effect that the global market will have on packaging.

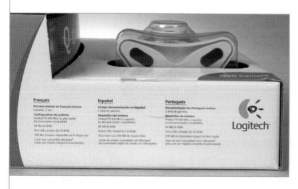

Growing global distribution of many products requires multiple languages on more and more retail packages

We are not even taking about language differences. It is true that many products sold globally already come in packages with multilingual text. On the American continent, many packages feature dual languages – English/French on packages sold in Canada, English/Spanish on many packages in U.S. Some packages feature all three languages used on the American continent.

On packages available in the EC area as many as eight or more languages are not uncommon. Hebrew and Arabic are standard in the Middle East. Japanese and other Asian languages can be found on packages marketed in the Far East. So, in the blossoming of global retail, multilingual packages are no longer unusual.

But less attention has been paid to the fact that future packaging needs to pay greater attention to *social* issues around the world. These relate to population growth and location, changing behavioral patterns, environmental issues, political issues, and many more.

The global population growth, sometimes estimated to be 13 billion people by mid-century, as well as economic initiatives, will put greater stress on distribution-related issues of packaged goods throughout the world. This, in turn, will make methods of transport, storage and distribution of packaged goods more prominent. Packages will have to be designed to prevent damage and deterioration in order to survive long periods of transportation, various weather conditions and dubious storage facilities at the point of delivery.

Take, for example, refrigeration. Americans joke about being served warm beer in the U.K. and lukewarm soft drinks in many countries. Ask for ice in your Coca-Cola in France and you will be lucky if a waiter will reluctantly bring you one little ice cube.

But it's no joke when you market products that *require* refrigeration, or even freezing, in countries where refrigeration facilities range from minimal to non-existent.

And then there is the difference in sales environments in many countries. In many of these countries, for example in Southeast Asia, Africa or South America, stores are frequently much smaller, even tiny and primitive in comparison to the stores in which we are accustomed to shop. Brand and product identification and text legibility have different demands. Marketers and designers in the Western world need to be familiar with these conditions so that their packages can function in these environments.

Also, more often than not, packages in these stores are not displayed on self-service shelves but stored behind counters, often stacked from floor to ceiling. There they are more difficult to see and their text is much more difficult, if not impossible, to read, calling for visual concepts that can communicate effectively under these conditions.

To facilitate the distribution of packaged goods among the low-income population in some of these areas, substantially different demands on package size and package cost must be considered.

Even in more affluent countries, where supermarkets and hypermarkets are not uncommon, try to find batteries bundled in units of 8, 12 and even 24 – a common sight in U.S. mass-merchandizing outlets. You will discover that two-unit blister packs are still the usual standard in many countries. There are parallels in many countries throughout the world where package size and unit count are much more modest than in the U.S. and other developed countries.

Old habits die slowly. Many people in Europe still prefer their daily visits at the local greengrocer, fishmonger and pharmacy, carrying their shopping bags home on foot or on their bicycle.

As global marketing becomes more prevalent, these issues will put particular emphasis on the need for exploring new materials and new ways of using these to package goods – whether it be food, medicine, clothing, housewares, electronics or any other product – and anticipating visual issues relating to global conditions,

Global environmental concerns

Then there is the issue of environmental concerns among a growing global community.

The relationship of industry to the environment is, as we indicated before, one topic that creates a lot of passionate debates all over the world. Packaging, already in the limelight – deservedly or not – as a major contributor to garbage proliferation, is not likely to escape this public scrutiny.

So, expect environmental issues to play a major role in future global packaging.

Some of the environmental concerns are not even a problem to which packaging contributes in a major way, even if environmental protagonists seize every opportunity to make it seem that way.

The changing climates throughout the world, brought on by global warming and the shrinking polar icecaps, may some day modify the production of some food and how it will be transported and stored at retail and at home. If this happens, marketers of such food and their package designers will need to anticipate these conditions and create or modify the design of relevant packaging, or develop new packaging concepts altogether.

The problem of global resources is another prominent issue. Deforestation and pollution, taking its toll in many of the world's areas, may lead marketers and packaging manufacturers, looking for alternative materials, to opt for replacing paper and paperboard with plastics, despite the disposal problems with plastic packages in many countries. Furthermore, if an increased need for plastic packaging does occur, this may depend on the availability of oil from countries whose political priorities may determine its availability – or unavailability – in some industrial countries.

All these controversial issues may lead to new, currently unknown materials, especially in the light of stressing the reduction of packaging materials and finding new ways of accomplishing this.

The development of new materials used for packaging will have to deal with the ever-present concerns about package disposal in many countries. Government rules for package disposal vary greatly from country to county, requiring global marketers and designers to keep constantly informed as to these requirements. For one thing, you can be sure that the vocal supporters of environmental issues will not let anyone forget their concerns about packaging materials and disposal problems.

These are among the many global issues that will influence the way packages will be designed and used in the future. And that will require visionary solutions by all participants in package development.

Internationally recognized symbols relating to environmental issues are used on packages throughout the world

Package designers can put a positive spin on this by being alert to opportunities for suggesting the visionary application of existing as well as new materials wherever and whenever possible.

Dealing with conflicting global demands

Despite the many social problems in various parts of the world, hardworking, two-income, mobile families are increasing in many countries and will demand packages that provide handling, storage and transportation features that cater to their different lifestyles.

Marketers, designers and packaging suppliers will need to come to terms with the tensions created by the conflicting demands of lifestyle, economic and environmental concerns.

Prepackaged food, more and more evident in huge quantities and varieties at hypermarkets, will become increasingly popular among mobile families. The current preference for plastic stock containers for these prepackaged meals, and the disposal problem that they represent, will call for explorations of new and visionary ways of packaging these products in terms of materials and structures.

Understanding the meaning of label information becomes more and more critical not only for the substantial aging population, but the overall population throughout the world.

Catering to the growing market of mobile consumers, prepackaged meals will demand many new visionary packaging concepts

But the ability of marketers, designers and suppliers to satisfy these demands is limited by the many conflicting regulations imposed by various national governments. This is a problem that government agencies in various countries need to address with greater vigor and better understanding of marketing complexities.

All these concerns present serious challenges to marketers, designers, material manufacturers and government agencies in the next few years. Participating parties can meet these challenges with either apprehension or exhilaration. With a positive approach, and working in coordination with each other, they present exciting opportunities for visionary solutions.

To our mind, this is the only way to go.

Chapter 22
The information bubble

Probably the most fascinating opportunities for visionary package design will be the availability of electronic devices able to accomplish all sorts of never previously performed tasks. The sky is the limit – literally.

In the first part of the 20th century, most packages were just containers to hold and store products.

In the middle of the 20th century, packages started becoming a communications medium, and brands had to compete against one another for your attention. This is when package designers started bringing their influence and offering advice on how one brand could outsell the other through good marketing.

The next generation has the package as an information-gathering device and a node on the Internet. This will be the revolutionary change in the next 20 years of packaging.

Let's consider the bar code that has appeared on every U.S. package for the past 30 years. It's been an excellent source of keeping track of products in the retail store and at the checkout counter. These bar codes are identical for every SKU (stock keeping unit). That means that each 16 oz Mott's Apple Sauce package across the world will use the same bar code.

But what if every individual product had its own coded information? Let's say billions of products around the world, each with its own code. This is the future, and the technical name for these individual codes or this type of information, is RFID (radio frequency identification).

These information devices actually got started in World War II, according to Kevin Ashton, executive director, Auto-ID Center at the MIT Media Laboratories, which is developing RFID for packaging. The Royal Air Force needed to make sure they didn't shoot down their own returning bombers and put automatic radio transmitters into their planes. It was a type of RFID.

Not long ago, however, several package goods companies and retailers realized that tracking their products efficiently and reducing inventory could save millions of dollars annually. The idea was to put a wireless network computer into each product, which would tell the story about how each product travels through the supply chain. When this idea first was suggested, it seemed outlandish, but the MIT Media Lab was willing to undertake the challenge of investigating it.

The technology today has reached the point of the silicon age, and the silicon chip will do the job. It's too expensive today for most packaging applications but, within the next few years, mass production could bring the price point down substantially. Wal-Mart has already started requiring its main suppliers to use RFID chips on their shipping cartons and pallets.

The price of silicon chips will go down with new technology and printing techniques, and at some point practically every package will have one.

What good is it?

We've had computers for 50 years, and the computers have become faster, smaller and cheaper. But the computer still doesn't know anything about the real world unless a human being feeds in the information. The real world is where we do business. The real world is where we live.

The RFID or smart microchip – a little bigger than a grain of sand – embedded in each package and shipping case will be a powerful tool. It will track real time inventory movement from the beginning of the supply chain to the end. Along with enabling extraordinary efficiency, some of the benefits of tagging packages worldwide with the chips may include:

- Store shelf replenishment.
- Information gathering.
- Product supply forcasting.
- Theft deterrence.
- Immense cost savings.

And hopefully, as a result, increased consumer satisfaction.

How long will your product be in the warehouse? How long will it be in the distribution center? Which store should that product be sent to? This is great promise for the future.

Remember how we said that computers can't do much on their own. These microchips, like little computers, have that problem, but they can do an enormous amount of work when talking to a larger computer.

Marketers want their products to leave the shelf in the hands of the consumer. You also want the shelves to be restocked as quickly as possible. These chips will automatically count what's on the shelf. The information on this tiny chip can be compared to a social security number. It's like a telephone network where you can dial a telephone number and talk to just about anybody. The number put on the chip – called an EPC (Electronic Product Code) – can be compared to a telephone number of the package.

So packages will no longer be only storage containers or marketing communicators, they will soon be a medium for networking worldwide. The Internet will work this technology. If a store needs to know when a particular product was made or was packaged, the store will scan the product, and using the Internet, communicate with the producer.

More free time

The information bubble will free up human beings to deal efficiently with their environments, according to Professor Sanjay Sarma at the MIT Media Laboratory. This is what Dr. Sarma and Dr. David Brock of the MIT Media Labs call "the philosophy of the information bubble". It is the overall reason for making packages more efficient in the future.

With RFID, when a package is on a shelf, you know where it is. The package is automatically reordered, so you have automatic shelf control, eventually automatic checkout, and even theft protection.

What are they doing at MIT to revolutionize packaging? First, they have to make the RFID chip so inexpensive that it can be used universally on packages. Second, they have to connect these RFIDs, not to just a reader, but to the world. The information stored on the chip could be connected through the internet or whatever large system can be used for information dissemination.

In fact, we someday will scan our products, hit a button, and have the new product delivered to you the next day, billed through the credit card.

Maybe those bottles of shampoo can be delivered more conveniently once you establish contact with the shampoo producers. Through RFIDs, the information put around the package makes it possible for you to make virtual contracts. This is a truly functional package. Your pretty bottles in the store will still attract attention, but once having established this "contract" with the shampoo producer, your bottles may be delivered in a more convenient manner.

These smart packages of the future need visionary designers who understand the technical issues that go with packaging, and also have a vision for how these technical issues can make the manufacturing process, distribution process and the consumer's life more simple.

Chapter 23
The digital package

Terrorism hit America's store shelves several years ago. Near Chicago someone had inserted cyanide into a few Tylenol bottles. This led to the recall of all Tylenol products for a while, and eventually to safety sealed packaging, which has become standard on many types of ingestible products.

If we had had RFID technology back then, we would have known where the product was made, distributed and purchased. Johnson & Johnson would have quickly been able to recall the products that needed to be recalled, rather than having to take the time and expense to pull out all the products nationwide.

Richard Cantwell, who now also consults with the Auto-ID Center at MIT Media Laboratories, had earlier in his career marketed Tylenol during that period of time. He's especially interested in the safety and security that RFID techonology can bring to products and, as a vice-president currently at Gillette, believes things like product pilferage will decrease substantially.

Brian Perkins, whose company, Johnson & Johnson, is also part of MIT's Auto-ID Center, talks about quality assurance. The smart packages will have a positive effect in the drug and pharmaceutical category. Think about compliance – your package can remind you when to take your pill, when it expires, when it's time to shop for another, and how to mix the product, if necessary. The new technology will also help accommodate legal information, along with the compliance information. Since baby boomers are making the pharmaceutical market grow, it's in for some booming years.

Pirating and stealing

We're not talking about the high seas. In marketing terms, pirating refers to making products that look like branded products so they can sell at high prices. It's like counterfeiting. In China pirating is prevalent, and people in garages make products that look like some of the brand names we're all familiar with, such as Tylenol, Lego, Polo and brands too numerable to name.

The counterfeiting problem is huge. For example, Gillette sees its razors and batteries counterfeited in certain parts of the world, and it now works with local police and the authorities to identify the counterfeiting sites. The next step is to get the authorities to shut down the cottage industries that are doing the counterfeiting. In the future, RFID will enable them to know whether the product is the authentic brand, and the retailers will be able to tell consumers confidently that the products are authentic.

Individual branded products with RFID will eventually be tracked at both the distribution level and the store level. Therefore, the potential for pirating authentic branded packaging will be reduced. This is a big breakthrough.

Pilferage has always been a problem with retail drug products. We all pay for pilferage when retailers lose money through products getting stolen or pilfered. The retailers add costs so the consumer actually ends up taking care of the losses.

Tamper-evident packaging might have prevented the Tylenol episode back then and it has become prevalent now in many drug and food products. Tamper-evident packages are even used for hardware and electronic products to prevent pilferage. Let's hope this doesn't raise the costs too high. Perhaps less pilferage will mean that the costs can stay within a reasonable level.

Retailers and manufacturers have for years discussed the advantages and disadvantages of electronic surveillance, since many items seem to walk out the door. The retailers tell the manufacturers to make their packages more pilfer-proof or they won't stock the packages. It was a big surprise when surveillance studies showed that half the pilferage is not from people taking the package out of the store, but from employees taking them.

What will the intelligent package do for the marketer? It will verify authenticity by tapping into instant and precise product information on when, where and by whom each item in the inventory was produced. And it will enable products to communicate in the home.

The digital package at home

Did you forget your daily medication? The smart medicine cabinet exists in prototype form in several pharmaceutical laboratories. Think about this. The medicine cabinet of the future reads the RFID chip on the package. It informs elderly people whether they've forgotten to take their daily pill, or whether the product has expired.

Here's an example:

You have a doctor's prescription filled at your local pharmacy. The label on the bottle tells you how often and how many tablets you should take each day. The label also indicates an expiry date, possibly six months from the time of purchase. You use the prescription for the determined period of time, after which you store the remaining tablets in your medicine cabinet for possible future use.

About a year later, you feel the need for the remedy again. Ah, good luck. You remember that you did not use all the tablets the first time around and that you kept the remaining ones in the medicine cabinet for just this kind of occasion.

You take a few tablets without being aware that they have passed their expiry date. This is especially likely if you are older and apt to be a little more forgetful.

In the future, the package will help you to remember. As you open the bottle, a little chip in the cap will signal to you, possibly by changing color or even by talking to you, reminding you that you need to renew the prescription, instead of using what you have.

Health insurers might even pay for this, because it keeps people healthier and cuts down costs. This and other business models can be part of a vision where the population is growing, as is the need for pharmaceuticals.

So we finally can get information of how to use our product. And at some point, these chips on the packages will provide cooking instructions to the microwave, or fabric care instructions to the washing machine, or life-saving dosage information to the medicine cabinet.

Chips embedded into the smart packages may change the way we shop, and make checkout easier at the supermarket, and easier at home to get information of how the product is best used. This is the convergence of computers and packaging. Sending instructions to your microwave, refrigerator and washing machine.

Does this make life easier, or more complicated? Visionary marketers can run with the opportunity. Knowing trends and lifestyles and what the future has in store can make our packages work better for us at home.

Another change in shopping will be the use of the package identity on the Internet, showing how packages can function on the Internet as an additional selling tool.

Negotiating with the consumer

Is Big Brother watching us? Will these chips give away everything we're doing? As consumers, we need to be assured that the information on how we're buying and using a product is safe and secure.

When we use our store card for so-called great discounts, we're also giving information to the store about what we purchase and ourselves. The RFID chips have the potential of giving a lot more information.

Marketers need to find ways to negotiate with consumers to get the information needed for a better, more efficient marketing message.

Kevin Ashton and his group at MIT hypothesized about a scenario where a refrigerator reads the chips, and everything we buy has chips in them. Your refrigerator communicates with your television, and your refrigerator says: "This guy drinks 12 liters of Diet Coke a week and he's about to run out." And the television network says: "Who wants to advertise to this guy? We have his favorite show, which he always watches, and we know he drinks a lot of Coke. Who wants to buy advertising time?" You can then imagine the bidding between soft drinks' suppliers to be the exclusive advertiser to this consumer during the show.

Hypothetical? Yes, but very valuable to marketers. Why waste money advertising to people who are unlikely to use your product?

Once the business world understands how to use this new information and uses it for a while, the information will move over to consumer usage. And that's when the consumer will have the technology in their home for automatic shopping list generation, washing machines that may sort out the wash, medicine cabinets that make sure you're taking the medication, and a host of other robotic-type applications.

What about refills?

A new opportunity for a growth market is "as-needed" packaging. The consumer buys one bottle of shampoo and the microchip indicates when the shampoo is almost used up. Refills arrive and are snapped into the original package. This will change the future of retailing. With a wide range of products and product categories, you'll have an automatic delivery system to which many consumers would subscribe. The retailer would become the source of the original product, and the manufacturer does the replenishment.

There is no reason why this type of replenishment shopping could not become global. There are cultural barriers to this, but as the U.S. becomes more efficient, the rest of the world will follow.

Perhaps the real revolution in packaging and retailing will be going from today's inanimate store shelf to tomorrow's market that figures out what's in your refrigerator and just mails to you what you need.

Has it expired?

When you buy a perishable product in a grocery store, supermarket or hypermarket, you look for the expiry date. You want to know how much time you have before the product is no longer safe to consume. Let's say the expiry date shown on the package is three weeks from today.

That sounds just fine to you. But what guarantee do you have that the package did not sit for several days in a un-air-conditioned warehouse, or an un-air-conditioned truck during transport from its point of origin?

Not to worry in the future. The little chip, embedded in the package will warn you when the product is spoiled, regardless of what the expiry date indicates.

The language puzzle

People around the world speak different languages and the packaging they buy accommodates it.

RFID can change the way your company markets its products. A central database could provide product information and save room on the package. This would allow the package designers to communicate the brand essence. All the language groupings can be placed on this central database, and consumers can gain access to the language that they speak.

The digital package in the store

How will retail change? Probably in an invisible manner. You won't notice, but the products will be there more often and there will be fewer people getting involved in bringing the products to the shelf.

One day you'll notice that the checkout counter is gone, and it's much easier to take your products out of the store without having to wait on long lines.

Consumers will appreciate that chips offer numerous opportunities to help in faster checkout, either at the checkout counter itself or even before you ever get to the checkout counter. You will be able to swipe your credit card or your store loyalty card across the package you have just put into your shopping cart so that, when you get to the checkout counter, your shopping list is already registered and your bill already waiting for you. All you have to do is sign for it and off you go.

Perhaps even the price of the product will change depending on how much of the product is in the supply chain.

There will be a somewhat invisible aspect to the retail revolution, invisible, that is, to the consumer. Unnoticed by the consumer, many changes will be occurring behind the scenes. Retailers will know when and how quickly products leave their inventory, increasing the accuracy with which products are made, shipped, delivered and stocked.

And new technology can enhance customer service by efficiently producing customized products on demand, managing special requests, and reducing time and costs to deliver goods to market.

When a product goes on the shelf, the store will immediately know whether it's in the right or wrong place. With the clerk's hand-held device, he will be alerted to return to a certain aisle where a product is out of stock or in the wrong department.

These shelves communicate with the package and tell the store and store personnel what needs to happen to make sure the consumer can find the product when they need it. This will dramatically change retailing.

Companies like Procter & Gamble and Unilever will be able to take advantage of RFID technology to make sure that their consumer products are properly organized on the shelf. And when popular fragrances, scents or sizes go out of stock, they will be replenished quickly. Remember, a major complaint retailers receive from consumers is that they went to the store to buy a product and could not find the product or size they were looking for.

So what is the major benefit of this new technology? Like any technology, we don't accurately know its major benefit yet.

Perhaps the biggest benefit will be the ability to keep shelves stocked and knowing when they're out of stock, and being able to identify where the merchandise is and alerting store personnel to get it out on shelf in time.

Retailers can also benefit when the smart shelf alerts the store if there is an unusual pattern of stock removal. This is one benefit, but the greater benefit is probably keeping the shelves stocked at all times.

Small, fast moving items, such as razor blades, toothbrushes or batteries, can fly off the shelf in peak periods. Cosmetics, such as lipsticks, need to be properly organized at all times. Store clerks don't always have the wherewithal or the time to keep an eye on shelves. When the chips in packages can communicate with the chips on the shelves, they are able to signal when they are empty and can thus be replenished quickly.

There are scams where consumers use receipts for products that they previously bought, then take an additional product off the shelf, bring it to customer service and ask for a refund. With RFID technology, you match the specific item to the receipt and customer service can tell if it was bought earlier at their store.

"Happy Birthday, John"

Can you imagine walking by a product in the store that you bought last year on your birthday, and it now speaks to you from the shelf? This sounds really far-fetched, but probably not impossible in the future.

Imagine a box that senses the consumer walking by it and can make that consumer more attracted to the box because of what the box knows about that consumer. You already have that level of consumer information at the website. You know about them from their purchasing patterns. You can put different marketing messages that are targeted to that consumer or group of consumers.

Next, you briefly stop at a shelf considering which of several brands to select. But what's this? A voice, activated by a chip in a package, talks to you, telling you about the benefits of the product in the package to induce you to buy it. As you touch a symbol on the package, the text on the package may change to address several factors about the product, or even change to a language with which you are most familiar.

You walk further down the isle and see a product you want. You wonder: Is the product on sale today? Your cellphone to the rescue! You point it at the package and not only find out whether it is on sale, but download all kinds of information about the product that there is not room enough on the package surfaces to provide. Or it could update information about promotional offers such as: "Now you can buy two of these products and we will give you XYZ product free."

The package may offer you the choice of selecting the language that you are most familiar with. Too far-fetched? Don't bet on it. Just look at all the kids holding up their cellphones to take pictures of their friends. Taking pictures with cellphones? Who would have thought this possible only a couple of years ago.

So it's not that big of a stretch to think that somewhere down the road, with RFID technology, a package could change its display or its message for a particular consumer. This might work off your cellphone or loyalty program card. Instead of scanning the card at the cash register, the package reads it as you walk by the display.

The brand and product information on the package might be the tip of the iceberg. But real information could be loaded within the computer chip that changes itself to attract the consumer.

The digital price strip

The shopper often has no idea how much money she will be paying at checkout, especially with stores that don't emphasize their pricing at the shelf. Compare your shopping receipt to what you thought the difficult-to-read price strip on the shelf told you, and you'll often find errors. This can change.

Some stores are now testing little pricing strips on fixtures, which are digitally controlled from a master control center inside retailers like Target, Wal-Mart or Best Buy. They're designed so that corporate headquarters can push a button at the start or end of the day and all 5,000 stores in two seconds have the price increase (or decrease). The digital marking on the shelf is prominent for consumers to see.

This digital pricing, digital ink and signage raises the initial investment cost. But the payout for retailers is great. According to Lee Carpenter, retailers in the future won't need as many people stocking shelves, changing shelves, changing specials, or going in to modify the price. Not only that, but it will be done accurately, and the signs will all be up to date and nicely designed so that the environment will look more professional.

The local store

Many of these larger retailers will spin out smaller versions. Wal-Mart, Best Buy, Publix Supermarkets and many others have already started the trend. Although the quantity of products sold within the smaller stores will be less, the strategy is to get into markets with smaller real estate. These smaller, more compact stores will accommodate people who want to shop locally.

Global retailers need to understand why countries use certain sizes of products, and how some markets sell more fresh products than others. The technology of electronic tracking will be used. This could show where there is a particular product or brand producing more returns. The retailer will be able to go back to the brand owner and say: "Some things are not working properly. Perhaps it's the packaging communication."

Stores and technology

Stores like Tesco in the U.K. are making large investments in what John Clarke refers to as "liberation technologies". These techniques liberate people from the back office and the checkout counter, and enable service people to work with the shopper – a better shopping experience. Tesco believes that a combination of this shopping experience and the technology to help it along will grow their business to a greater level.

For example, sales assistants use PDA-type devices which can tell shoppers immediately whether a product is available and where it's located. The mission for the sales assistants at Tesco is to make sure the shopper is happy and relieved of long queues.

Tesco is a big proponent of the Auto-ID Center at MIT and will be using its RFID chip for more insights into trends and package replenishment. Originally, the chips will go into the packages and labels for pricier products such as appliances and clothing.

Tesco is evaluating how technology can help the consumer to make a selection in the store and locate things more easily. One vision it has is for consumers to use their mobile phone as a shopping device to help them shop for clothing or food. For example, if they want to find a certain food product, it's entered through your phone key – when you're near the product, perhaps the tone will rise. Or, perhaps if you're looking for jeans, you tap your size in your phone key and it will point you to the nearest pair.

As you are a club cardholder, Tesco knows what you buy. If you want a healthy, low fat diet, it'll send the information to you. You then swipe your cell phone over the product, your cellphone reads the RFID chip and tells you if it's the right product for you. According to Mr. Clarke, this is more than just packaging and technology. It's consumer insight – helping the customer make better choices.

The privacy puzzle

As you read about RFID's positive qualities, you might also start to worry about your privacy. According to the respected publication the *Guardian*:

> Retailers have hailed RFID as the "Holy Grail" of supply chain management, but some civil liberties groups argue that they are "spy chips" and an invasion of consumers' privacy. They even argue that the chips could be used as a covert surveillance device.

If personal identity were linked with RFID, could you be profiled and tracked without your knowledge and consent? This is a roadblock to RFID's widespread use that must be considered. Privacy advocates worry about the ramifications of embedding the chips into all kinds of personal items.

Spokespersons from MIT's Auto-ID Center say they're taking precautions to protect consumer privacy, such as calling for retailers to be able to disable the tags after the product is purchased. But this "kill" feature is still in its test stage.

For the near future, consumer products with RFID will have the tags embedded in the package rather than the products, so the chips can be disposed of.

Many suggestions have been offered towards protecting your privacy. CNET News, on the Internet, has closely followed the progress of this new technology. Declan McCullagh, CNET's political correspondent, indicates that the tags are fine for tracking pallets and crates, but if retailers wish to use them on consumer goods, he recommends voluntary guidelines:

- Consumers should be notified – perhaps on the checkout receipt – when RFID tags are present on their purchase.
- The tags should be disabled by default at the checkout counter.
- The tags should be placed on the package instead of the product.
- The tags should be readily visible and easily removable.

The large privacy rights group CASPIAN (Consumers Against Supermarket Privacy Invasion and Numbering) goes even further. It wants manufacturers and retailers to agree to a voluntary moratorium on the item-level RFID tagging of consumer items until a formal technology assessment process involving all stakeholders, including consumers, can take place.

It wants the development of the technology to be guided by a strong set of 'Principles of Fair Information Practices', ensuring that meaningful consumer control is built into the implementation of RFID technology.

Finally, it says that some uses of RFID technology are inappropriate in a free society and should be flatly prohibited: Merchants should be prohibited from forcing or coercing customers into accepting live RFID tags in their products, and there should be no prohibition on individuals to detect and disable the tags on items in their possession. RFID tags must not be used to track individuals without their informed consent, since human tracking is inappropriate. And RFID should not be employed to eliminate or reduce anonymity, such as incorporating it into currency.

When today's familiar bar codes were first introduced in the early 1970s, there were also consumer concerns. But the manufacturer, retailer and consumer know they have to work together to keep one another happy and stay in business, so eventually the RFID "rules of the road" will enable the new technology to benefit everyone. As we indicated earlier in this chapter, along with RFID you need the vision to negotiate with the consumer to get the information needed for a better, more efficient marketing message.

Business-to-business

A food market in South Africa will communicate with a market in California when trading partners have access to the same global product information code. This is an exciting challenge in retail and packaging, and is described as "electronic collaboration". E-collaboration connects the entire retail industry globally through one common data and information-sharing system. The idea is to enable industries to conduct transactions in real time using connected databases.

A group called UCC.net, a subsidiary of the Uniform Code Council, has an Internet-based forum where data synchronization and data registry occurs. This allows all user companies, no matter how small or where they are located, to match up their e-commerce language.

Together with the RFID, or smart, microchip, e-collaboration can combine the knowledge of available products with tracing inventory movement.

The smart tag readers on store shelves, and in warehouses and trucks, will transmit information to a data bank, creating the business-to-business efficiencies we've spoken about.

Will every package on the planet be turned in to a computer? Very possibly.

There will even come a time when packaging will sense things to do with its contents. Perhaps 20 years from now we'll be able to sense bacteria in food through these microchips.

Packages as electronic information-gathering devices will continue to evolve radically. Having a unique number for each packaged product eventually enables you to have infinite information. Businesses will look for efficiencies, cost savings, labor reductions and faster supply chains. They'll gain information that they never had before about how their business functions.

This enables you to change the way you bring your products to market. With vision, you'll understand your consumer better than your competitor, and bring upgrades and enhancements that your consumer wants.

Sure, packaging five years and ten years from now might look quite similar to today's packaging, but its role in the business process and the things it is capable of doing will be radically different. It will be gathering information and communicating information. What is the temperature of my soda? How long will this perishable food last?

Companies spend fortunes today on individuals counting their products and replenishing the shelves. This will become automatic.

Will the store change?

Have you been to stores like Toys R Us and Babies R Us where you get a scanner and can go around scanning different baby items that you want to register for gifts? Friends and relatives can view your selected items on the Internet, order them and have them delivered to your doorstep. Basically the store is reduced to simply a physical place for you to look at things and make selections, not necessarily to purchase them and take them home.

Scanning the packages you like at Toys R Us gives friends and relatives the option of viewing and buying the products through the Internet

Many store chains are looking for ways to save their customers' time from what some consider the chore of shopping. The hand-held scanner, used by the shopper, promises them faster checkout in the future.

In a test in the Chicago area, the loyalty card members of the Jewel-Osco supermarket chain can shop, scan and bag their own groceries.

The Jewel-Osco hand-held scanner also accesses the shoppers' buying history, and even flashes coupons on the screen as they walk past shelves containing items that they might be prone to purchase. In fact, these loyalty card members get discounts not necessarily available to regular shoppers.

The scanners also show a running total of the shoppers' carts as they buy. And the scanners can be hooked up to self-checkout aisles, so the shopper can move along at a quick pace.

Shopping in the store of the future, according to Professor Sarma at MIT, could theoretically mean going once a month, scanning items, and deciding which should be delivered, and when, from your home over the next several weeks.

The technical challenge is here with packaging and retailing. We need the vision to find the proper balance of technology with personal service. Good retailers will play up their strengths, which to most shoppers means good service, and will not use technology to sterilize their environment. This is your technical challenge.

Chapter 24
New challenges for marketers and package designers

All this is not only a challenge for retailers, consumers and those who develop these electronic components, but marketers and package designers will have to adjust to working and living with these new devices. It's a good bet that, in a few years hence, electronic components are certain to become an integral ingredient of virtually every package.

Brand and product managers will have to learn to evaluate how and when to use the electronic devices on their packages, and for what purpose, with regard to criteria, appropriateness, effectiveness and cost.

To take advantage of creating "smart packages" by incorporating electronic devices, either hidden or deliberately displayed, as part of their design assignments, package designers will need to acquire an intimate understanding of the use and some of the technical aspects of the RFID systems. They will need to understand how to apply them, how they function, and the cost implications of incorporating them.

But will this love affair with digital gadgets cut into and outweigh previously critical packaging issues, such as brand identification and product glamorization? Or will the popularity of e-commerce, buying merchandise on the Internet, instead of at the brick-and-mortar store, have the effect of minimizing packaging, relegating it to a minor role in the marketing of retail products?

The package designer's concern

In the increasingly competitive retail world, the growing interest by marketers of reaching consumers by experimenting with new methods of merchandising should surprise no one. And so, the introduction of new electronic devices in packaging becomes a tantalizing experiment that could change our shopping habits in many, as yet unpredictable, ways.

But as with anything new, there is always, as the saying goes, a fly in the ointment. And so it is with this new intrusion of digital devices into the world of packaging.

As we are all aware, during the 1990s, e-commerce had a cataclysmic impact on every type of business. It came out of nowhere and, for a while, charged ahead with fierce determination, threatening to take control of virtually all retail business.

New and seemingly dynamic business ventures, each with their own brand identification, appeared virtually daily on our monitor screens. Investors, eager to take advantage of various get-rich-quick promises, flocked to support these fledglings.

With all this flurry of electronic activities, retail marketers and people involved in various segments of the packaging industry began to wonder out loud whether and how packaging would be affected by this technological juggernaut.

As more and more enterprises joined the cyber-world of marketing, marketers and designers began to wonder how packaging would fare in this new, and still virgin, territory. Will buying products on the Internet, they wondered, change the role of packaging? What role, if any, will packaging play? Will viewing product promotions on a monitor screen minimize the role of packaging? Will packaging be as necessary and integral to buying products as in the brick-and-mortar environment?

Is not some merchandise already offered on the Internet without showing any packages, or by displaying packages so minutely that they are barely recognizable, let alone readable? Should marketer and designers be prepared for the potential of "package-less" retailing on the Internet?

There seemed, indeed, ominous signs on the horizon. Alarm bells started ringing in market management and design offices. Package design agencies became especially nervous. Would packaging lose its position as an important sales tool? Could packaging become an insignificant detail in online marketing?

Could packaging – perish the thought – just fade away?

Package design – Who cares?

The package design community took these concerns very seriously. They were well aware that the proliferation of website offerings of every conceivable type of merchandise without showing packaging, or showing infinitesimally small packaging images, could leave its mark on the effectiveness of the packaging medium.

A seminar under the title of "Package Design – Who Cares?" featured several well-known professionals who addressed concerns as to where packaging, and package design specifically, was heading in the light of the website phenomenon. While there was no consensus among the speakers, the preoccupation as to the future of packaging among the predominantly designer audience was clearly visible.

Branding online and offline

By now, it is clear that the alarm about the possible disappearance of packaging has been premature, if not essentially unfounded. E-commerce, even at its peak, has had little impact on packaging function or package appearance.

It is, of course, possible that the continued interest in e-commerce could still lead to changes in marketing that might eventually affect package design even if, at this point of time, the roller-coaster ride that e-commerce has experienced makes prophecies of how and in what manner very uncertain.

For the time being, the emergence of e-commerce has brought to the surface one important distinction. From here on, packaging will have the double duty of having to function effectively both online and offline.

For the time being, at least, packages purchased online and delivered at home are unlikely to be any different from those acquired in brick-and-mortar stores. Consumers expect to find the same reassuring brand identity, the same information and the same visual appearance of the packages to which they have been accustomed when purchasing the products at the neighborhood store.

But the jury is still out

The age of electronic communications has only started. It is certain to continue and grow, even if at temporarily reduced speed. But as marketing methods change, packaging will have to keep in step with marketing requirements, or packaging *will* fade away as a major marketing component.

So it should not surprise anyone that, as time goes by, the state of e-commerce will have *some* residual effect on the way packages will look in the next decade. Meanwhile, marketers, packaging suppliers and package designers will have to live with the uncertainty of anticipating some kind of fallout from this powerful electronic sales medium.

There could be changes in the size and shapes of packages or what kind of materials will be used for them. It may alter the way brand names are used, or the amount and size of text on packages, or the way visuals are treated.

Adjustments in package colors may be needed to improve clear Internet transmission. Packages shown in TV ads are almost always simplified or modified to enhance legibility and color reproduction on the TV screen. Packages in online product promotions may need to be treated in a similar manner to appear at their best on the monitor screen.

This will need to be continuously explored by marketers and designers. Designers will have to crank up their visionary genius to explore how to maximize packaging effectiveness on the Internet. It may make the difference between a product sold and a product bypassed.

What will not change in all this confusing guesswork are the basic elements that packages for all products require: good visibility online and offline, strong brand identification, distinct brand imagery and product perception, clear and comprehensible copy and, last but not least, emotional appeal.

For shoppers, especially the growing number of young, mobile shoppers, to be excited about your products, you will need to stretch your imagination about how to present your products in packaging that will anticipate their emotional response to your brand and your products.

What should be the role of package design in this?

Chapter 25
On your mark – get ready – get set – GO!

I have the simplest of tastes … I am always satisfied with the best.

Oscar Wilde

The next decade, you can be sure, will be a tumultuous decade – at times exciting and at other times infuriating.

When tastes and lifestyles change and many sources of merchandise are targeting an audience that is becoming younger, more discriminating and more mobile every day, new discoveries, innovations, opportunities – and, yes, new obstacles – are bound to be a daily occurrence in any business.

To meet these challenges requires a proactive, precisely targeted, yet flexible attitude in every segment of the marketing cycle, along with comprehension of and sensitivity to the needs and desires of the fast multiplying young generation that will dominate the next decade.

If that doesn't smell like change is in the air, what does?

No problem. We're in luck!

Targeting the young generation during the next decade will be balanced by the availability of more and more young people taking on responsibilities previously held by their older, more conservative predecessors. Less encumbered by marketing traditions than their forerunners and pressured by economic urgencies as well as their personal ambitions, they will seek unique ways of marketing merchandise. They will know how to appeal to the young because they are young themselves.

But wait a minute! Are we leaving something out?

What about all those reports we read about the generation of seniors multiplying fast as a result of more sophisticated medical services and constantly improving pharmaceutical products? Who says that targeting young consumers will be the best way, the only way, to go?

And what about the increasing emergence of ethnic groups in various countries – Hispanics in the U.S., Eastern European, Middle Eastern and African immigrants into Western Europe – each of them contributing their own particular cultural traits and differing needs?

To keep pace with all these marketing realities, marketers will need to understand this cultural melting pot or acquire a special insight of one or the other of these cultural communities. In either event, to take a more holistic approach to strategic development, the matrix management approach, that calls for various experts to sit around the table to communicate with each other, will become the desirable means of achieving creative design solutions.

At the same time, design agencies that want to participate in their clients' strategic development, will need to become accustomed to dealing with a younger, more educated and more sophisticated marketing clientele who will look for design agencies that show an in-depth understanding of their clients' business categories.

With competitive pressures increasing, design agencies will be expected to participate in developing brand strategy as an integral part of their brand identity and package design assignments, provided that they can demonstrate that the principals and their staff have the professional experience and willingness to become intimately involved in their clients' strategic needs.

To create visionary packaging, meetings regarding brand strategies must include the design agency from the beginning and throughout the brand development

On the other side of the table, we again emphasize that for this to happen, marketers must have an open mind to include their design agency in participating actively in every step of strategy development, from pre-design research, meetings regarding strategic and financial planning, post-design consumer research, coordination with their pro-

duction and purchasing departments and involvement in package final-
ization and production procedures.

It is entirely possible that this may not only change the method in
which design agencies and their clients work together, but it may affect
methods of payment. These may resemble the financial arrangements
between advertising agencies and their clients, or retainer arrangements that both designers and marketers, each for their own reasons,
have previously been reluctant to enter.

Undoubtedly, veteran design consultants, many of them already
experienced in brand strategy development, will be eager to participate
in a broader scope of working relationship with their clients, to the
benefit of both parties.

But there is a price to pay. With their attention diverted to strategic
issues and away from "the board", many of the responsibilities for
visual creation, a skill to which design principals personally contributed for many years, will have to be shifted to equally competent
staff associates.

Filling the creative gap

We're in luck again! The influx of many young designers in the brand
identity and package design profession during recent years has created
a substantial pool of highly motivated practitioners.

As the principals of design agencies are turning their attention to
helping their clients in a wide scope of marketing and brand-related
issues, young designers will fill the gap of doing the creative work.
Having been born with a computer virtually in their cribs, they have
turned the design business upside down. They are fiercely motivated,
highly creative and technically conversant.

Make no mistake about it. Computers have dramatically changed
the whole universe of the design profession. Computers have the
ability to expand the designer's skills to include areas that, a few
years ago, were the sole territory of specialists. There was a time
when comp renderers, illustrators, calligraphers, typographers,
retouchers, photostat operators and mechanical artists had thriving
careers assisting design agencies and studios by freelancing for them
or being in their employ.

Today's designers look quizzical at the mention of these methods, as
if they were remnants of the Stone Age, although it is barely a dozen
years since the computer seriously took over the majority of design
development. But the new breed of designers, having grown up

knowing no other means of design creation, are not hesitant to use computers aggressively.

Colors, once limited to available color papers, watercolors or Magic Markers applied to paperboard pinned to the drafting table, now can be imported liberally and in great variety on the monitor and changed or modified instantly. Typography, formerly favoring such venerable classic forms as Helvetica or Garamond, has been joined by such a variety of type styles that it boggles the imagination and they can be stretched, altered, curved or whatever, in minutes.

Having the computer at their disposal, designers now have total control over all these activities. Their ability enables them to incorporate previously separate skills – setting type, creating logos, importing illustrations from various sources, changing design elements and colors at will, enlarging and reducing them with a few clicks on the keyboard. Instead of having to rely on the capabilities, availability and delivery schedules of external services, they now have more time to experiment and react more quickly to clients' delivery requirements.

The Internet even enables designers to gather information, insights and ideas from all over the world.

Add to this the amazing progress being made in reproduction technology. Smaller, digitally operated and functioning printing equipment suggests the possibility of in-house production for small print runs, or even packaging to order, bringing the control of package design and package production full cycle.

Most importantly, today's young creative lions are more adventurous and less encumbered by the design hang-ups of their older peers. Instead of playing safe, they are willing to take some risks.

New faces on the horizon

Meanwhile, the choice of design practitioners is widening for those who seek a totally new approach to their branding and package design requirement. New faces in the design community are appearing on the horizon. Designers who are noted for their achievements in one particular discipline are crossing the line to new areas of creativity.

In fact, in the next decade, having the reputation as a source for creative inventiveness in a broad field may become more important than being identified as a specialist in a particular design discipline. It's really not that novel. Did not architects, such as Frank Lloyd Wright and Michael Graves, successfully design everything from furniture to textiles and household goods, in addition to houses?

The next decade will require designers and design agencies to take a much more holistic approach. Clients will expect them to offer creative input in a number of categories in which they were not previously involved. These may cover as wide a range as the development of brand identity and the design of ads, brochures, annual reports, retail environments and, of course, packaging. Some design agencies are gearing up for just such an expected mix of services.

What will dominate the design service industry in the future is the ability to think creatively, to be inspirational, to reach beyond the confines of popular mediocrity in a wide range of disciplines that require taste and professional experience.

We can see it already today.

Fashion designers are being approached by marketers to apply their creative talents to totally different industries, to create visual programs for everything from museum exhibits to airline environments. Design agencies that have been primarily active in graphics for annual reports, advertising and promotional campaigns suddenly find themselves exploring new and unexpected concepts for branding programs and packaging.

More than the invasion of online shopping and other recent commercial developments, diversity in creative expectation will become the primary challenge for marketers and designers, as well as for educational facilities that teach marketing and design, in the next ten years.

Most significantly, it opens up new vistas of opportunities for marketing and brand managers whose creative appetite is sufficiently tempted to try previously untraveled roads.

It's a different world out there

It's a baby boomer mentality. It's the challenge to be different. It's fun to be different. It's a chance to be visionary. It's an opportunity to take advantage of these talented young designers, even at the risk that some of them may occasionally be overdoing their creative freedom.

Your designers have grown up in an environment that tends to be creatively more experimental. The computer has provided them with a tool that has virtually no limitations. Being able to access an immense range of images with relative ease, their whole way of looking at design has expanded their ability to create packaging concepts with greater flair and imagination.

There is, of course, always a chance that designers will occasionally fall off the cliff in their effort to be different, to be "creative". But look at it this way: It will be the price well worth paying for the opportunity that their effort will spark a totally unexpected solution. A visionary solution? Perhaps. You can never know. Remember the old saying: Nothing ventured, nothing gained.

Business in the twenty-first century will be driven by an ever-increasing hunger for new creations and new developments. New packaging concepts will offer unique opportunities for those marketers who are willing to take the initiative and an occasional well-planned risk. Those willing to accept this premise will be the most likely to prevail in the crowded, play safe market.

No doubt about it. Marketing strategies in the next five or ten years will require an attitude of GET READY – GET SET – *GO!*

Tomorrow's young brand and product managers will be entrusted with greater marketing responsibilities than ever before. They will be more flexible – perhaps at times a little recklessly – in dealing with more unconventional concepts of packaging than their predecessors. Even as management will hold them accountable for their decisions, they will be less reluctant than their predecessors to take risks and will be receptive to fresh and unique ideas.

The bottom line is the promise of greater opportunities, greater vision, greater success.

For packaging to succeed in this kind of market environment will require the courage and willingness by both management and their designers to be proactive and to take occasional risks.

Risks, not gambles. Risks based on solid, carefully crafted criteria. Risks that will be productive and visionary. Risks that will captivate consumers. Risks that will create market leaders.

As the French playwright, Pierre Corneille, said in his play *Le Cid*:

"To conquer without risk is to triumph without glory."

Index

Quality, of bottles 108
Quilmes 145–6
Qumran 11

Radio frequency
  identification (RFID)
  216–28
  background 213–14
  the future 191, 215
  in the home 144, 218–19,
  220
  information-gathering 219
  medical uses 218
  privacy and 219, 225–6
  safety issues 216
  in the store 142, 221–3
  tracking products 217
refills 219–20
refrigeration, global
  marketing and 209
regulations 189–91, 212
replenishment shopping
  220
research
  bringing opportunities
  195–7
  into branding 77–8
  online 196
  qualitative 196
  quantitative 196
retail stores
  diminishing numbers of
  staff 133, 207, 221, 223
  emotional element 127
  the future 207
  importance of packaging
  3
  as marketing medium 109
  mergers 199
  needs of 120–1
  private labels 111
  store brand development
  117
  store environment
  109–28
  see also local stores;
  supermarkets
RFID, see radio frequency
  identification
risk-taking 51, 52, 238
  product innovation 196–7

Safety 204
sales environments, national
  209–10
sales staff, diminishing

numbers 133, 207, 221,
  223
scams, and RFID
  technology 222
scanners, for customers
  227–8
self-dispensing packs 93–4
sensory reaction 48–9
shapes of packages
  96–108
  Absolut vodka bottles 96,
  137–8
  of bottles 27–8, 69–70,
  99–100
  as brand identifiers 44
  continuing improvements
  96–7
  ketchup 27–8
  liquor bottles 75
  paint containers 104
  shaped cans 18, 61
shipping cases/containers
  62, 63–4, 122–3
  at Wal-Mart 63, 121,
  133–4
shopping
  impulse 109–10, 121
  rational 66
shopping bags, sensory
  response 48–9
shopping experience 126–7
  help for shoppers 123–4
simplicity 79, 121
  in shopping 80
smart labels 142, 151, 153
smart microchips, see radio
  frequency identification
smart packages 229
smart tag readers 227
Smucker's 79
snacks 83–4
social issues, in global
  marketing 208–10
soft drinks 18
sophistication 125
soup, see Campbell's Soups
special effects 141–2
spray bottles 91–2
Stamp Act 1765 14
Starkist 103
sterilization 15, 16
stock control 222
stores, see local stores;
  retail stores; supermarkets
styrene foam 19

supermarkets
  arrangement in 81–2
  mergers 199
  owning local stores 224
  private labels 23–4, 35–6,
  36–7, 111
  proliferation of, after WWII
  20, 21
  replacing local stores 20,
  23
  see also retail stores
surveillance, electronic 217

Tabasco Sauce 74
tag readers, smart 227
tamper evidence 202, 217
Target Stores 65–6, 113–14
tea, iced 84–5
technology
  in stores 221, 224–5
  see also radio frequency
  identification
technology products 73
Teddy Graham 83–4
television
  commercials 109
  product placement
  139–41
  zapping commercials 139
temperature, global aspects
  121–2
Tesco 112, 123–4
  home deliveries 124
  Internet shopping 124
  label system 36
  liberation technologies
  224–5
  shopping experience 126
theft, see pilferage
tinfoil 18
tin plating 15, 16
toiletries, men's 72–4
Tostitos 173–4
Toys R Us, scanners 227–8
tracking, electronic 224
trade dress 204–5
  legal protection of 29
Trader Joe's 115
Ts'ai Lun 13
tuna fish 102–3
typography, computers and
  236

UCC.net 181, 226
UPS 151–3

delivery status via Internet
  152
  sorting for delivery 153

Vases, ancient 8–9
vending machines 105–6
vinegars 82–3
violators 201–2
virtual management
  179–88
  advertising 187
  benefits of 185–6
  leaving the system 187–8
  offshoot of the brand
  consultant 180, 184–5
  offshoot of the
  printer/converter 180,
  183–4
  offshoot of the product
  catalog 180–3
  systems 180–5
vision 1–3
  definition 1
visual difficulties, older
  people 55–6
visuals on packages 47

Wal-Mart 129–35
  brands 114
  diabetes care products
  132
  efficiency 132–3
  moving pallets 134
  packaging 63–4, 129–30
  pharmacy department
  131–2
  prices 129, 130–1
  price/value relationship
  132–3, 134
  shelf stacking 133–4
  shipping cases/containers
  63, 121, 133–4
  value 130
warnings, on packages 191
water, bottled 76–7
web-based knowledge
  management, see virtual
  management
Wegmans 126, 127
Weight Watchers 120
wine 85–7
wine vinegars 82–3

Yogurt 76